T0145223

Lecture Notes in Social Networks

Lecture Notes in Social Networks (LNSN) comprises volumes covering the theory, foundations and applications of the new emerging multidisciplinary field of social networks analysis and mining. LNSN publishes peer-reviewed works (including monographs, edited works) in the analytical, technical as well as the organizational side of social computing, social networks, network sciences, graph theory, sociology, Semantics Web, Web applications and analytics, information networks, theoretical physics, modeling, security, crisis and risk management, and other related disciplines. The volumes are guest-edited by experts in a specific domain. This series is indexed by DBLP. Springer and the Series Editors welcome book ideas from authors. Potential authors who wish to submit a book proposal should contact Christoph Baumann, Publishing Editor, Springer e-mail: Christoph.Baumann@springer.com

More information about this series at http://www.springer.com/series/8768

Kai Shu • Suhang Wang • Dongwon Lee • Huan Liu
Editors

Disinformation, Misinformation, and Fake News in Social Media

Emerging Research Challenges
and Opportunities

 Springer

Editors
Kai Shu
Computer Science & Engineering
Arizona State University
Tempe, AZ, USA

Suhang Wang
College of Information Sciences
and Technology
The Pennsylvania State University
University Park, PA, USA

Dongwon Lee
College of Information Sciences
and Technology
The Pennsylvania State University
University Park, PA, USA

Huan Liu
Computer Science & Engineering
Arizona State University
Tempe, AZ, USA

ISSN 2190-5428   ISSN 2190-5436   (electronic)
Lecture Notes in Social Networks
ISBN 978-3-030-42701-6   ISBN 978-3-030-42699-6   (eBook)
https://doi.org/10.1007/978-3-030-42699-6

This Springer imprint is published by the registered company Springer Nature Switzerland AG.
The registered company address is: Gewerbestrasse 11, 6330 Cham, Switzerland

Acknowledgment

We wish to express our profound gratitude to the contributing authors of the chapters of the book: without their expertise and dedicated efforts, this book would not have been possible. We are grateful to Christoph Baumann and Christopher Coughlin who provided guidance and assistance on the publishing side of this effort. Last but not the least, we are deeply indebted to our families for their support throughout this entire project. We dedicate this book to them, with love.

Tempe, AZ, USA   Kai Shu
University Park, PA, USA   Suhang Wang
University Park, PA, USA   Dongwon Lee
Tempe, AZ, USA   Huan Liu

Contents

Mining Disinformation and Fake News: Concepts, Methods, and Recent Advancements

Kai Shu, Suhang Wang, Dongwon Lee, and Huan Liu

Abstract In recent years, disinformation including fake news, has became a global phenomenon due to its explosive growth, particularly on social media. The wide spread of disinformation and fake news can cause detrimental societal effects. Despite the recent progress in detecting disinformation and fake news, it is still non-trivial due to its complexity, diversity, multi-modality, and costs of fact-checking or annotation. The goal of this chapter is to pave the way for appreciating the challenges and advancements via: (1) introducing the types of information disorder on social media and examine their differences and connections; (2) describing important and emerging tasks to combat disinformation for characterization, detection and attribution; and (3) discussing a weak supervision approach to detect disinformation with limited labeled data. We then provide an overview of the chapters in this book that represent the recent advancements in three related parts: (1) user engagements in the dissemination of information disorder; (2) techniques on detecting and mitigating disinformation; and (3) trending issues such as ethics, blockchain, clickbaits, etc. We hope this book to be a convenient entry point for researchers, practitioners, and students to understand the problems and challenges, learn state-of-the-art solutions for their specific needs, and quickly identify new research problems in their domains.

Keywords Disinformation · Fake news · Weak social supervision · Social media mining · Misinformation

K. Shu (✉) · H. Liu
Computer Science and Engineering, Arizona State University, Tempe, AZ, USA
e-mail: kai.shu@asu.edu; dongwon@psu.edu

S. Wang · D. Lee
College of Information Sciences and Technology, The Penn State University, University Park, PA, USA
e-mail: huan.liu@asu.edu; szw494@psu.edu

© Springer Nature Switzerland AG 2020
K. Shu et al. (eds.), *Disinformation, Misinformation, and Fake News in Social Media*,
Lecture Notes in Social Networks, https://doi.org/10.1007/978-3-030-42699-6_1

1

Social media has become a popular means for information seeking and news consumption. Because it has low barriers to provide and disseminate news online faster and easier through social media, large amounts of disinformation such as fake news, i.e., those news articles with intentionally false information, are produced online for a variety of purposes, ranging from financial to political gains. We take fake news as an example of disinformation. The extensive spread of fake news can have severe negative impacts on individuals and society. First, fake news can impact readers' confidence in the news ecosystem. For example, in many cases the most popular fake news has been more popular and widely spread on Facebook than mainstream news during the U.S. 2016 presidential election.[1] Second, fake news intentionally persuades consumers to accept biased or false beliefs for political or financial gain. For example, in 2013, $130 billion in stock value was wiped out in a matter of minutes following an Associated Press (AP) tweet about an explosion that injured Barack Obama.[2] AP said its Twitter account was hacked. Third, fake news changes the way people interpret and respond to real news, impeding their abilities to differentiate what is true from what is not. Therefore, it is critical to understand how fake news propagate, developing data mining techniques for efficient and accurate fake news detection and intervene to mitigate the negative effects.

This book aims to bring together researchers, practitioners and social media providers for understanding propagation, improving detection and mitigation of disinformation and fake news in social media. Next, we start with different types of information disorder.

1 Information Disorder

Information disorder has been an important issue and attracts increasing attention in recent years. The openness and anonymity of social media makes it convenient for users to share and exchange information, but also makes it vulnerable to nefarious activities. Though the spread of misinformation and disinformation has been studied in journalism, the openness of social networking platforms, combined with the potential for automation, facilitates the information disorder to rapidly propagate to massive numbers of people, which brings about unprecedented challenges. In general, information disorder can be categorized into three major types: disinformation, misinformation, and malinformation [1]. *Disinformation* is fake or inaccurate information that is intentionally spread to mislead and/or deceive. *Misinformation* is false content shared by a person who does not realize it is false or misleading. *Malinformation* is to describe genuine information that is shared with an intent

---

[1]https://www.buzzfeednews.com/article/craigsilverman/viral-fake-election-news-outperformed-real-news-on-facebook

[2]https://www.telegraph.co.uk/finance/markets/10013768/Bogus-AP-tweet-about-explosion-at-the-White-House-wipes-billions-off-US-markets.html

to cause harm. In addition, there are some other related types of information disorder [2, 3]: *rumor* is a story circulating from person to person, of which the truth is unverified or doubtful. Rumors usually arise in the presence of ambiguous or threatening events. When its statement is proved to be false, a rumor is a type of misinformation; *Urban Legend* is a fictional story that contains themes related to local popular culture. The statement and story of an urban legend are usually false. An urban legend is usually describing unusual, humorous, or horrible events; *Spam* is unsolicited messages sent to a large number of recipients, containing irrelevant or inappropriate information, which is unwanted.

The spread of false or misleading information often has a dynamic nature, causing the exchanging among different types of information disorder. On the one hand, disinformation can become misinformation. For example, a disinformation creator can intentionally distribute the false information on social media platforms. People who see the information may be unaware that it is false and share it in their communities, using their own framing. On the other hand, misinformation can also be transformed into disinformation. For example, a piece of satire news may be intentionally distributed out of the context to mislead consumers. A typical example of disinformation is fake news. We use it as a tangible case study to illustrate the issues and challenges of mining disinformation (Fig. 1).

1.1 Fake News as an Example of Disinformation

In this subsection, we show how disinformation (fake news) can be characterized, detected, and attributed with social media data. Fake news is commonly referred as the news article that are intentionally and verifiably false and could mislead readers [4, 5].

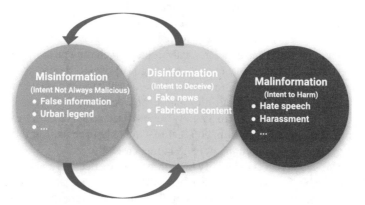

Fig. 1 The illustration of the relations among disinformation, misinformation, and malinformation, with representative examples (e.g., fake news is an example of disinformation). In addition, misinformation and disinformation can be converted mutually

For ***characterization***, the goal is to understand whether the information is malicious, has harmless intents, or has other insightful traits. When people create and distribute disinformation they typically have a specific purpose in mind, or intent. For example, there can be many possible intents behind the deception including: (1) persuade people to support individuals, groups, ideas, or future actions; (2) persuade people to oppose individuals, groups, ideas or future actions; (3) produce emotional reactions (fear, anger or joy) toward some individual, group, idea or future action in the hope of promoting support or opposition; (4) educate (e.g., about vaccination threat); (5) prevent an embarrassing or criminal act from being believed; (6) exaggerate the seriousness of something said or done (e.g., use of personal email by government officials); (7) create confusion over past incidents and activities (e.g., did the U.S. really land on the moon or just in a desert on earth?); or (8) demonstrate the importance of detecting disinformation to social platforms (e.g., Elizabeth Warren and Mark Zuckerberg dispute). End to end models augmented with feature embeddings such as causal relations between claims and evidence can be used [6] to detect the intents such as Persuasive influence detection [7]. Once we have identified the intent behind a deceptive news article, we can further understand how successful this intent will be: what is the likelihood that this intent will be successful in achieving its intended purpose. We can consider measures of virality grounded in social theories to aid characterization. Social psychology points to social influence (how widely the news article has been spread) and self-influence (what preexisting knowledge a user has) as viable proxies for drivers of disinformation dissemination [8]. Greater influence from the society and oneself skews a user's perception and behavior to trust a news article and to unintentionally engage in its dissemination. Computational social network analysis [9] can be used to study how social influence affects behaviors and/or beliefs of individuals exposed to disinformation and fake news.

When the entire news ecosystem is considered instead of individual consumption patterns, social dynamics emerge that contribute to disinformation proliferation. According to social homophily theory, social media users tend to follow friend like-minded people and thus receive news promoting their existing narratives, resulting in an echo chamber effect. To obtain a fine-grained analysis, we can treat propagation networks in a hierarchical structure, including macro-level such as posting, reposting, and micro-level such replying [10], which shows that structural and temporal features within information hierarchical propagation networks are statistically different between disinformation and real news. This can provide characterization complementary to a purely intent-based perspective, for instance to amplify prioritization of disinformation that may quickly have undesirable impacts after being shared with benign intent (e.g., humor) initially.

For ***detection***, the goal is to identify false information effectively, at a early stage, or with explainable factors. Since fake news attempts to spread false claims in news content, the most straightforward means of detecting it is to check the truthfulness of major claims in a news article to decide the news veracity. Fake news detection on traditional news media mainly relies on exploring news content information. News content can have multiple modalities such as text, image, video. Research has explored different approaches to learn features from single or combined modalities and build machine learning models to detect fake news. In addition to features related directly to the content of the news articles, additional social context features can be derived from the user-driven social engagements of news consumption on social media platform. Social engagements represent the news proliferation process over time, which provides useful auxiliary information to infer the veracity of news articles. Generally, there are three major aspects of the social media context that we want to represent: users, generated posts, and networks. First, fake news pieces are likely to be created and spread by non-human accounts, such as social bots or cyborgs. Thus, capturing users' profiles and behaviors by user-based features can provide useful information for fake news detection [11]. Second, people express their emotions or opinions toward fake news through social media posts, such as skeptical opinions and sensational reactions. Thus, it is reasonable to extract post-based features to help find potential fake news via reactions from the general public as expressed in posts. Third, users form different types of networks on social media in terms of interests, topics, and relations. Moreover, fake news dissemination processes tend to form an echo chamber cycle, highlighting the value of extracting network-based features to detect fake news.

Fake news often contains multi-modality information including text, images, videos, etc. Thus, exploiting multi-modality information has great potentials to improve the detection performance. First, existing work focuses on extracting linguistic features such as lexical features, lexicon, sentiment and readability for binary classification, or learning neural language features with neural network structures, such as convolution neural networks (CNNs) and recurrent neural networks (RNNs) [12]. Second, visual cues are extracted mainly from visual statistical features, visual content features, and neural visual features [13]. Visual statistical features represent the statistics attached to fake/real news pieces. Visual content features indicate the factors describing the content of images such as clarity, coherence, diversity, etc. Neural visual features are learned through neural networks such as CNNs. In addition, recent advances aim to extract visual scene graph from images to discover common sense knowledge [14], which greatly improve structured scene graphs from visual content.

For *attribution*, the goal is to verify the purported source or provider and the associated attribution evidence. Attribution search in social media is a new problem because social media lacks a centralized authority or mechanism that can store and certify provenance of a piece of social media data. From a network diffusion perspective, identify the provenance is to find a set of key nodes such that the information propagation is maximized [9]. Identifying provenance paths can indirectly find the originated provenances. The provenance paths of information are usually unknown, and for disinformation and misinformation in social media it is still an open problem. The provenance paths delineate how information propagates from the sources to other nodes along the way, including those responsible for retransmitting information through intermediaries. One can utilize the characteristics of social to trace back to the source [15]. Based on the Degree Propensity and Closeness Propensity hypotheses [16], the nodes with higher degree centralities that are closer to the nodes are more likely to be transmitters. Hence, it is estimated that top transmitters from the given set of potential provenance nodes through graph optimization. We plan to develop new algorithms which can incorporate information other than the network structure such as the node attributes and temporal information to better discover provenances.

With the success of deep learning especially deep generative models, machine-generated text can be a new type of fake news that is fluent, readable, and catchy, which brings about new attribution sources. For example, benefiting from the adversarial training, a series of language generation models are proposed such as SeqGAN [17], MaliGAN [18], LeakGAN [19], MaskGAN [20], etc. and unsupervised models based on Transformer [21] using multi-task learning are proposed for language generation such as GPT-2 [22] and Grover [23]. One important problem is to consider machine-generated synthetic text and propose solutions to differentiate which models are used to generate these text. One can perform classification on different text generation algorithms' data and explore the decision boundaries. The collections of data can be acquired from representative language generation models such as VAE, SeqGAN, TextGAN, MaliGAN, GPT-2, Grover, etc. In addition, meta-learning can be utilized to predict new text generation sources from few training examples. Moreover, some generative models such as SentiGAN [24], Ctrl [25] and PPLM [26], can generate stylized text which encodes specific styles such as emotional and catchy styles. It is important to eliminate spurious correlations in the prediction model, e.g., disentangling style factors from the synthetic text using adversarial learning, and develop prediction models with capacity to recover transferable features among different text generation models.

2   The Power of Weak Social Supervision

Social media enables users to be connected and interact with anyone, anywhere and anytime, which also allows researchers to observe human behaviors in an unprecedented scale with new lens. User engagements over information such as news articles, including posting about, commenting on or recommending the news on social media, bear implicit judgments of the users to the news and could serve as sources of labels for disinformation and fake news detection.

However, significantly different from traditional data, social media data is big, incomplete, noisy, unstructured, with abundant social relations. This new (but weak) type of data mandates new computational analysis approaches that combine social theories and statistical data mining techniques. Due to the nature of social media engagements, we term these signals as *weak social supervision* (WSS). We can learn with weak social supervision to understand and detect disinformation and fake news more effectively, with explainability, at an early stage, etc. Generally, there are three major aspects of the social media engagements: users, contents, and relations (see Fig. 2). First, users exhibit different characteristics that indicate different patterns of behaviors. Second, users express their opinions and emotions

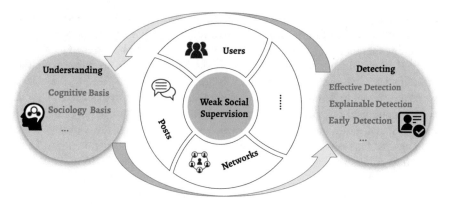

Fig. 2 The illustration of learning with social supervision for understanding and detecting disinformation and fake news

through posts/comments. Third, users form different types of relations on social media through various communities. The goal of weak social supervision is to leverage signals from social media engagements to obtain weak supervision for various downstream tasks. Similar to weak supervision, we can utilize weak social supervision in the forms of weak labels and constraints.

2.1 Understanding Disinformation with WSS

Humans are not naturally good at differentiating misinformation and disinformation. Several cognitive theories explain this phenomenon, such as *Naïve Realism* and *Confirmation Bias*. Disinformation mainly targets consumers by exploiting the individual vulnerabilities of news consumers. With these cognitive biases, disinformation such as fake news is often perceived as real. Humans' vulnerability to fake news has been the subject of interdisciplinary research, and these results inform the creation of increasingly effective detection algorithms. To understand the influence of disinformation and fake news in social media, we can employ techniques to characterize the dissemination from various types of WSS: (1) sources (credibility/reliability, trust, stance/worldview, intentions) [27, 28]; (2) targeted social group (biases, demographic, stance/worldview) [11]; (3) content characteristics (linguistic, visual, contextual, emotional tone and density, length and coherence) [5, 8]; and (4) nature of their interactions with their network (e.g., cohesive, separate) [9]. For example, the effects of these theories can be quantified by measuring user metadata [11], to answer the question "why people are susceptible to fake news?", or "Are specific groups of people more susceptible to certain types of fake news?".

Some social theories such as social identity theory suggests that the preference for social acceptance and affirmation is essential to a person's identity and self-esteem, making users likely to choose "socially safe" options when consuming and disseminating news information. According to social homophily theory, users on social media tend to follow and friend like-minded people and thus receive news promoting their existing narratives, resulting in an echo chamber effect. Quantitative analysis is a valuable tool for verifying whether, how, and to what magnitude these theories are predictive of user's reactions to fake news. In [29], the authors made an attempt to demonstrate that structural and temporal perspectives within the news hierarchical propagation networks can affect fake news consumption, which indicates that additional sources of weak social supervision are valuable in the fight against fake news. To obtain a fine-grained analysis, propagation networks are treated in a hierarchical structure, including macro-level (in the form of posting, reposting) and micro-level (in the form of replying) propagation networks. It is observed that the features of hierarchical propagation networks are statistically different between fake news and real news from the structural, temporal and linguistic perspectives.

2.2 Detecting Disinformation with WSS

Detecting disinformation and fake news poses unique challenges that makes it non-trivial. First, the *data challenge* has been a major roadblock because the content of fake news and disinformation is rather diverse in terms of topics, styles and media platforms; and fake news attempts to distort truth with diverse linguistic styles while simultaneously mocking true news. Thus, obtaining annotated fake news data is non-scalable, and data-specific embedding methods are not sufficient for fake news detection with little labeled data. Second, the *evolving challenge* of disinformation and fake news, meaning, fake news is usually related to newly emerging, time-critical events, which may not have been properly verified by existing knowledge bases (KB) due to the lack of corroborating evidence or claims. To tackle these unique challenges, we can learn with *weak social supervision* for detecting disinformation and fake news in different challenging scenarios such as *effective*, *explainable*, and *early* detection strategies. The outcomes of these algorithms provide solutions to detecting fake news, also provide insights to help researchers and practitioners interpret prediction results.

Effective detection of disinformation The goal is to leverage weak social supervision as an auxiliary information to perform disinformation detection effectively. As an example, interaction networks are used for modeling the entities and their relationships during news spreading process to detect disinformation. Interaction networks describe the relationships among different entities such as publishers, news pieces, and users (see Fig. 3). Given the interaction networks the goal is to embed the different types of entities into the same latent space, by modeling the interactions among them. The resultant feature representations of news can be leveraged to perform disinformation detection, with the framework <u>Tri</u>-relationship for <u>F</u>ake <u>N</u>ews detection (TriFN) [30].

Fig. 3 The TriFN model of learning with social supervision from publisher bias and user credibility for effective disinformation detection [30]

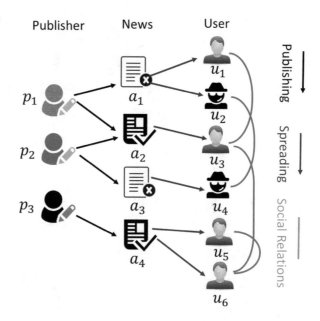

Inspired from sociology and cognitive theories, the weak social supervision rules are derived. For example, social science research has demonstrated the following observations which serves our weak social supervision: *people tend to form relationships with like-minded friends, rather than with users who have opposing preferences and interests.* Thus, connected users are more likely to share similar latent interests in news pieces. In addition, for publishing relationship, the following weak social supervision can be explored: *publishers with a high degree of political bias are more likely to publish disinformation.* Moreover, for the spreading relation, we have: *users with low credibilities are more likely to spread disinformation, while users with high credibility scores are less likely to spread disinformation.* Techniques such as nonnegative matrix factorization (NMF) is used to learn the news representations by encoding the weak social supervision. Experiments on real world datasets demonstrate that TriFN can achieve 0.87 accuracy for detecting disinformation.

Confirming disinformation with explanation Take fake news as an example, explainable disinformation detection aims to obtain top-k explainable news sentences and user comments for disinformation detection. It has the potential to improve detection performance and the interpretability of detection results, particularly for end-users not familiar with machine learning methods. It is observed that not all sentences in news contents are fake, and in fact, many sentences are true but only for supporting wrong claim sentences. Thus, news sentences may not be equally important in determining and explaining whether a piece of news is fake or not. Similarly, user comments may contain relevant information about the important aspects that explain why a piece of news is fake, while they may

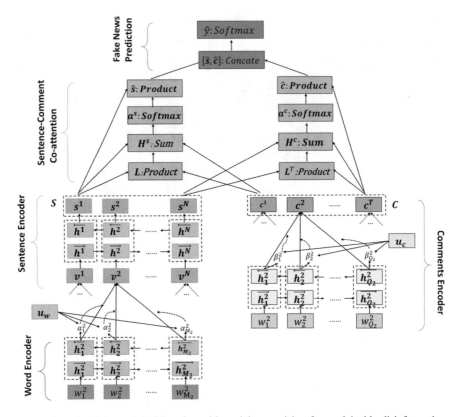

Fig. 4 The dEFEND model of learning with social supervision for explainable disinformation detection [31]

also be less informative and noisy. The following weak social supervision can be used: *the user comments that are related to the content of original news pieces are helpful to detect fake news and explain prediction results.* In [31], it first uses Bidirectional LSTM with attention to learn sentence and comment representations, and then utilizes a sentence-comment co-attention neural network framework called dEFEND (see Fig. 4) to exploit both news content and user comments to jointly capture explainable factors. Experiments show that dEFEND achieves very high performances in terms of accuracy (~ 0.9) and F1 (~ 0.92). In addition, dEFEND can discover explainable comments that improve the exaplainability of the prediction results.

Early warning for disinformation Disinformation such as fake news is often related to newly emerging, time-critical events, which may not have been verified by existing knowledge bases or sites due to the lack of corroborating evidence. Moreover, detecting disinformation at an early stage requires the prediction models to utilize minimal information from user engagements because extensive user engagements indicate more users are already affected by disinformation. Social media data is multi-faceted, indicating multiple and heterogeneous relationships between news pieces and the spreaders on social media. First, users' posts and comments have rich crowd information including opinions, stances, and sentiment that are useful to detect fake news. Previous work has shown that conflicting sentiments among the spreaders may indicate a high probability of fake news [32, 33]. Second, different users have different credibility levels. Recent studies have shown some less-credible users are more likely to spread fake news [30]. These findings from social media have great potential to bring additional signals to early detection of fake news. Thus, we can utilize and learn with multi-source of weak social supervision simultaneously (in the form of weak labels) from social media to advance early fake news detection.

The key idea is that in the model training phase, social context information is used to define weak rules for obtaining weak labeled instances, in addition to the limited clean labels, to help training. In the prediction phase (as shown in Fig. 5), for any news piece in test data, only the news content is needed and no social engagements is needed at all, and thus fake news can be detected at a very early stage. A deep neural network framework can be used where the lower layers of the network learn *shared* feature representations of the news articles, and the upper layers of the network *separately* model the mappings from the feature representations to each of the different sources of supervision. The framework MWSS aims to exploit jointly Multiple sources of Weak Social Supervision besides the clean labels. To extract the weal labels, the following aspects are considered including sentiment, bias, and credibility.

First, research has shown that news with conflicting viewpoints or sentiments is more likely to be fake news [32]. Similarly, it has been shown that use opinions

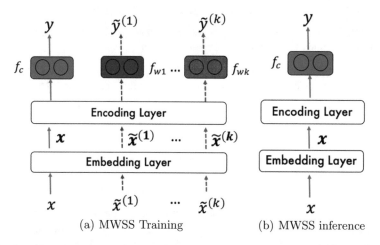

(a) MWSS Training   (b) MWSS inference

Fig. 5 The MWSS framework for learning with multiple weak supervision from social media data for early detection of disinformation. (**a**) During training: it jointly learns clean labels and multiple weak sources; (**b**) During inference, MWSS uses the learned feature representation and function f_c to predict labels for (unseen) instances in the test data

towards fake news have more sentiment polarity and less likely to be neutral [34]. Thus, the sentiment scores are measured (using a widely used tool VADER [35]) for all the users spreading the news, and then measure the variance and desperation of the sentiment scores by computing their standard deviation. We have the following weak labeling function:

> **Sentiment**-based: If a news piece has an standard deviation of user sentiment scores greater than a threshold τ_1, then the news is weakly labeled as fake news.

Second, social studies have theorized the correlation between the bias of news publisher and the veracity of news pieces [36]. To some extent, the users who post the news can be a proxy of the publishers and their degree of bias toward fake news and real news is different [11]. Thus, the news with users who are more biased are more likely to share fake news, and less biased users are more likely to share real news. Specifically, method in [37] is adopted to measure user bias cores by exploiting users' interests over her historical tweets. The bias score lies in range $[-1, 1]$ where -1 indicates left-leaning and $+1$ means right-leaning. We have the following weak labeling function:

> **Bias**-based: If the average value of user absolute bias scores is greater than threshold τ_2, then the news pieces is weakly-labeled as fake news.

Third, studies have shown that those less credible users, such as malicious accounts or normal users who are vulnerable to fake news, are more likely to spread fake news [30, 33]. The credibility score means the quality of being trustworthy of the user. To measure user credibility scores,we adopt the practical approach in [27]. The basic idea in [27] is that less credible users are more likely to coordinate with each other and form big clusters, while more credible users are likely to from small clusters. Note that the credibility score is inferred from users' historical contents on social media. We have the following weak labeling function:

> **Credibility**-based: If a news piece has an average credibility score less than a threshold τ_3, then the news is weakly-labeled as fake news.

The proper threshold is decided with a held-out validation dataset. Experimental results show that MWSS can significantly improve the fake news detection performance even with limited labeled data.

3   Recent Advancements: An Overview of Chapter Topics

In this section, we demonstrate the recent advancements of mining disinformation and fake news. This book is composed of three parts, and we give an overview of the chapter topics as follows. Part I consists of 5 chapters (2 to 6) on understanding the dissemination of information disorder. Part II contains 4 chapters (7 to 10) on techniques for detecting and mitigating disinformation, fake news, and misinformation. Part III includes 4 chapters (11 to 14) on trending issues such as ethics, block chain, clickbaits.

Part I: User Engagements in the Dissemination of Information Disorder

- Understanding the characteristics of users who are likely to spread fake news is an essential step to prevent gullible users to deceived by disinformation and fake news. In the Chap. 2, it presents a content-based approach to predict the social identity of users in social media. This chapter first introduces a self-attention hierarchical neural network for classifying user identities, and then demonstrates its effectiveness in a standard supervised learning setting. In addition, it shows good performance in a transfer learning setting where the framework is first trained in the coarse-grained source domain and then fine-tuned in a fine-grained target domain.

- User engagements related to digital threats such as misinformation, disinformation, and fake news are important dimensions for understanding and potentially defending against the wide propagation of digital threats. The Chap. 3 performs a quantitative comparison study to showcase the user characteristics of different user groupings who engage in trustworthy information, disinformation and misinformation. It aims to answer the following questions: (1) who engage with (mis)information and (dis)information; (2) what kind of user feedback do user provide; and (3) how quickly do users engage with (mis) and (dis)information? The empirical results to these questions indicates the clear differences of user engagement patterns, which potentially help the early handling of misinformation and disinformation.

- Understanding disinformation across countries is important to reveal the essential factors or players of disinformation. In the Chap. 4, the authors propose to identify and characterize malicious online users into multiple categories across countries of China and Russia. It first performs a comparison analysis on the differences in terms of networks, history of accounts, geography of accounts, and bot analysis. Then it explores the similarity of key actors across datasets to reveal the common characteristics of the users.

- In the Chap. 5, the authors study the misinformation in the entertainment domain. This chapter compares two misinformation-fueled boycott campaigns through examination of their origins, the actors involved, and their discussion over time.

- While many users on social media are legitimate, social media users may also be malicious, and in some cases are not even real humans. The low cost of creating social media accounts also encourages malicious user accounts, such as social bots, cyborg users, and trolls. The Chap. 6 presents an overview the use of bots to manipulate the political discourse. It first illustrates the definition, creation and detection of bots. Then it uses three case studies to demonstrate the bot characteristics and engagements for information manipulation.

Part II: Techniques on Detecting and Mitigating Disinformation

- Limited labeled data is becoming the largest bottleneck for supervised learning systems. This is especially the case for many real-world tasks where large scale annotated examples can be too expensive to acquire. In the Chap. 7, it proposes to detect fake news and misinformation using semi-supervised learning. This chapter first presents three different tensor-based embeddings to model content-based information of news articles which decomposition of these tensor-based models produce concise representations of spatial context. Then it demonstrate a propagation based approach for semi-supervised classification of news articles when there is scarcity of labels.

- With the development of multimedia technology, fake news attempts to utilize multimedia content with images or videos to attract and mislead consumers for rapid dissemination, which makes visual content an important part of fake news. Despite the importance of visual content, our understanding about the role of visual content in fake news detection is still limited. The Chap. 8 presents a comprehensive review of the visual content in fake news, including the basic concepts, effective visual features, representative detection methods and challenging issues of multimedia fake news detection. It first presents an overview of different ways to extract visual features, and then discusses the models including content-based, knowledge-based approaches. This chapter can help readers to understand the role of visual content in fake news detection, and effectively utilize visual content to assist in detecting multimedia fake news.

- The Chap. 9 proposes to model credibility from various perspectives for fake news detection. First, it presents how to represent credibility from the sources of the news pieces in authors and co-authors. Then it extracts signals from news content to represent credibility including sentiments, domain expertise, argumentation, readability, characteristics, words, sentences, and typos. Finally, these credibility features are combined and utilized for fake news detection and achieve good performances in real world datasets.

- The intervention of misinformation and disinformation is an important task for mitigating their detrimental effects. In the Chap. 10, the authors propose two frameworks to intervene the spread of fake news and misinformation by increasing the guardians' engagement in fact-checking activities. First, it demonstrates how to perform personalized recommendation of fact-checking articles to mitigate fake news. Then it tries to generate synthetic text to increase the engagement speed of fact-checking content.

Part III: Trending Issues

- In the Chap. 11, it focuses on the evaluation of fake news literacy. This chapter first introduce social media information literacy (SMIL) in general. Then it applies SMIL into the context of fake news including semantic characteristics, emotional response and news sources. Finally, this chapter discusses several promising directions for both researchers and practitioners.

- The Chap. 12 presents a dataset, AI system, and browser extension for tackling the problem of incongruent news headlines. First, incongruent headlines are labor-intensive to annotate, the chapter proposes an automatically way to generate datasets with labels. Second, it proposes a deep neural network model that contains a hierarchical encoder to learn the representations for headlines for prediction, and demonstrate the effectiveness in real world datasets. Finally, a web interface is developed for identifying incongruent headlines in practice.

- The Chap. 13 presents an overview of the evolving YouTube information environment during the NATO Trident Juncture 2018 exercise, and identifies how commenters propel video's popularity while potentially shaping human behavior through perception. This research reveals effective communication strategies that are often overlooked but highly effective to gain tempo and increase legitimacy in the overall information environment.
- The Chap. 14 presents a comprehensive survey on using blockchain technology to defend against misinformation. First, it gives the definition and basic concepts of blockchain. Second, it discusses how blockchain can be utilized to combat misinformation with representative approaches. Moreover, this chapter points out several promising future directions on leveraging for fighting misinformation.

4 Looking Ahead

Fake news and disinformation are emerging research areas and have open issues that are important but have not been addressed (or thoroughly addressed) in current studies. We briefly describe representative future directions as follows.

4.1 Explanatory Methods

In recent years, computational detection of fake news has been producing some promising early results. However, there is a critical piece of the study, the explainability of such detection, i.e., why a particular piece of news is detected as fake. Recent approaches try to obtain explanation factors from user comments [31] and web documents [38]. Other types of user engagements such as user profiles can be also modeled to enhance the explainability. In addition, explaining why people are gullible to fake news and spread it is another critical task. One way to tackle this problem is from a causal discovery perspective by inferring the directed acyclic graph (DAG) and further estimate the treatment variables of users and their spreading actions.

4.2 Neural Fake News Generation and Detection

Fake news has been an important problem on social media and is amplified by the powerful deep learning models due to their power of generating neural fake news [23]. In terms of neural fake news generation, recent progress allows malicious users to generate fake news based on limited information. Models like Generative Adversarial Network (GAN) [19] can generate long readable text from noise and GPT-2 [22] can write news stories and fiction books with simple context. Existing fake news generation approaches may not be able to produce style-enhanced and fact-enriched text, which preserves the emotional/catchy styles and relevant topics related to news claims. Detecting these neural fake news pieces firstly requires us to understand the characteristic of these news pieces and detection difficulty. Dirk Hovy et al. propose an adversarial setting in detecting the generated reviews [39]. [23] and [40] propose neural generation detectors that fine-tune classifiers on generator's previous checkpoint. It is important and interesting to explore: (i) how to generate fake news with neural generative models? (ii) can we differentiate human-generated and machine-generated fake/real news?

4.3   Early Detection of Disinformation

Detecting disinformation and fake news at an early stage is desired to prevent a large amount of people to be affected. Most of the previous work learns how to extract features and build machine learning models from news content and social context to detect fake news, which generally considers the standard scenario of binary classification. More recent work consider the setting that few or even no user engagements are utilized for predicting fake news. For example, Qian et al. propose to generate synthetic user engagements to help the detection of fake news [41]; Wang et al. present an event-invariant neural network model to learn transferable features to predict newly whether emerging news pieces are fake or not. We also discussed how we can utilize various types of WSS to perform early detection of fake news in Sect. 2.2. We can enhance these techniques with more sophisticated approaches that rely on less training data, for instance few-shot learning [42] for early fake news detection.

4.4   Cross Topics Modeling on Disinformation

The content of fake news has been shown to be rather diverse in terms of topics, styles and media platforms [33]. For a real-world fake news detection system, it is often unrealistic to obtain abundant labeled data for every domain (e.g., Entertainments and Politics are two different domains) due to the expensive labeling cost. As such, fake news detection is commonly performed in the single-domain setting, and supervised [43] or unsupervised methods [44, 45] are proposed to handle limited or even unlabeled domains. However, the performance is largely limited due to overfitting on small labeled samples or without any supervision information. In addition, models learned on one domain may be biased and might not perform well on a different target domain. One way to tackle this problem is to utilize domain adaptation techniques to explore the auxiliary information to transfer the knowledge from the source domain to the target domain. In addition, advanced machine learning strategies such as adversarial learning can be utilized to further capture the topic-invariant feature representation to better detect newly coming disinformation.

Acknowledgements This material is based upon work supported by, or in part by, ONR N00014-17-1-2605, N000141812108, NSF grants #1742702, #1820609, #1915801. This work has been inspired by Dr. Rebecca Goolsby's vision on social bots and disinformation via interdisciplinary research.

References

1. Wardle, C., Derakhshan, H.: Information disorder: toward an interdisciplinary framework for research and policy making. Council of Europe Report, 27 (2017)
2. Wu, L., Morstatter, F., Carley, K.M., Liu, H.: Misinformation in social media: definition, manipulation, and detection. ACM SIGKDD Explor. Newsletter **21**(2), 80–90 (2019)
3. Zhou, X., Zafarani, R.: Fake news: a survey of research, detection methods, and opportunities. arXiv preprint arXiv:1812.00315 (2018)
4. Tandoc, E.C. Jr., Lim, Z.W., Ling, R.: Defining "fake news" a typology of scholarly definitions. Digit. Journal. **6**(2), 137–153 (2018)
5. Shu, K., Liu, H.: Detecting Fake News on Social Media. Synthesis Lectures on Data Mining and Knowledge Discovery. Morgan & Claypool Publishers, San Rafael (2019)
6. Hidey, C., McKeown, K.: Identifying causal relations using parallel wikipedia articles. In: Proceedings of the 54th Annual Meeting of the Association for Computational Linguistics (Volume 1: Long Papers), pp. 1424–1433 (2016)
7. Hidey, C.T., McKeown, K.: Persuasive influence detection: the role of argument sequencing. In: Thirty-Second AAAI Conference on Artificial Intelligence (2018)
8. Zhou, X., Zafarani, R., Shu, K., Liu, H.: Fake news: fundamental theories, detection strategies and challenges. In: WSDM (2019)
9. Shu, K., Bernard, H.R., Liu, H.: Studying fake news via network analysis: detection and mitigation. CoRR, abs/1804.10233 (2018)
10. Shu, K., Mahudeswaran, D., Wang, S., Liu, H.: Hierarchical propagation networks for fake news detection: investigation and exploitation. In: ICWSM (2020)
11. Shu, K., Wang, S., Liu, H.: Understanding user profiles on social media for fake news detection. In: MIPR (2018)
12. Oshikawa, R., Qian, J., Wang, W.Y.: A survey on natural language processing for fake news detection. arXiv preprint arXiv:1811.00770 (2018)
13. Cao, J., Guo, J., Li, X., Jin, Z., Guo, H., Li, J.: Automatic rumor detection on microblogs: a survey. arXiv preprint arXiv:1807.03505 (2018)
14. Bosselut, A., Rashkin, H., Sap, M., Malaviya, C., Celikyilmaz, A., Choi, Y.: Comet: commonsense transformers for automatic knowledge graph construction. arXiv preprint arXiv:1906.05317 (2019)
15. Gundecha, P., Feng, Z., Liu, H.: Seeking provenance of information using social media. In: Proceedings of the 22nd ACM International Conference on Information & Knowledge Management, pp. 1691–1696. ACM (2013)
16. Barbier, G., Feng, Z., Gundecha, P., Liu, H.: Provenance data in social media. Synthesis Lectures on Data Mining and Knowledge Discovery **4**(1), 1–84 (2013)
17. Yu, L., Zhang, W., Wang, J., Yu, Y.: SeqGAN: sequence generative adversarial nets with policy gradient. In: Thirty-First AAAI Conference on Artificial Intelligence (2017)
18. Che, T., Li, Y., Zhang, R., Hjelm, R.D., Li, W., Song, Y., Bengio, Y.: Maximum-likelihood augmented discrete generative adversarial networks. arXiv preprint arXiv:1702.07983 (2017)

19. Guo, J., Lu, S., Cai, H., Zhang, W., Yu, Y., Wang, J.: Long text generation via adversarial training with leaked information. In: Thirty-Second AAAI Conference on Artificial Intelligence (2018)
20. Fedus, W., Goodfellow, I., Dai, A.M.: Maskgan: better text generation via filling in the_. arXiv preprint arXiv:1801.07736 (2018)
21. Vaswani, A., Shazeer, N., Parmar, N., Uszkoreit, J., Jones, L., Gomez, A.N., Kaiser, L., Polosukhin, I.: Attention is all you need (2017)
22. Radford, A., Wu, J., Child, R., Luan, D., Amodei, D., Sutskever, I.: Language models are unsupervised multitask learners. OpenAI Blog **1**(8), 9 (2019)
23. Zellers, R., Holtzman, A., Rashkin, H., Bisk, Y., Farhadi, A., Roesner, F., Choi, Y.: Defending against neural fake news. arXiv preprint arXiv:1905.12616 (2019)
24. Wang, K., Wan, X.: Sentigan: generating sentimental texts via mixture adversarial networks. In: IJCAI (2018)
25. Keskar, N.S., McCann, B., Varshney, L.R., Xiong, C., Socher, R.: Ctrl: a conditional transformer language model for controllable generation. arXiv preprint arXiv:1909.05858 (2019)
26. Dathathri, S., Madotto, A., Lan, J., Hung, J, Frank, E., Molino, P., Yosinski, J., Liu, R.: Plug and play language models: a simple approach to controlled text generation. arXiv preprint arXiv:1912.02164 (2019)
27. Abbasi, M.-A., Liu, H.: Measuring user credibility in social media. In: SBP (2013)
28. Tang, J., Liu, H.: Trust in social computing. In: Proceedings of the 23rd International Conference on World Wide Web, pp. 207–208. ACM (2014)
29. Shu, K., Mahudeswaran, D., Wang, S., Liu, H.: Hierarchical propagation networks for fake news detection: investigation and exploitation. In: ICWSM (2020)
30. Shu, K., Wang, S., Liu, H.: Beyond news contents: the role of social context for fake news detection. In: WSDM (2019)
31. Shu, K., Cui, L., Wang, S., Lee, D., Liu, H.: Defend: explainable fake news detection. In: ACM KDD (2019)
32. Jin, Z., Cao, J., Zhang, Y., Luo, J.: News verification by exploiting conflicting social viewpoints in microblogs. In: AAAI (2016)
33. Shu, K., Sliva, A., Wang, S., Tang, J., Liu, H.: Fake news detection on social media: a data mining perspective. KDD Explor. Newsletter (2017)
34. Cui, L., Lee, S.W.D.: Same: sentiment-aware multi-modal embedding for detecting fake news (2019)
35. Hutto, C.J., Gilbert, E.: Vader: a parsimonious rule-based model for sentiment analysis of social media text. In: ICWSM (2014)
36. Gentzkow, M., Shapiro, J.M., Stone, D.F.: Media bias in the marketplace: theory. In: Handbook of Media Economics, vol. 1, pp. 623–645. Elsevier (2015)
37. Kulshrestha, J., Eslami, M., Messias, J., Zafar, M.B., Ghosh, S., Gummadi, K.P., Karahalios, K.: Quantifying search bias: investigating sources of bias for political searches in social media. In: CSCW. ACM (2017)
38. Popat, K., Mukherjee, S., Yates, A., Weikum, G.: Declare: debunking fake news and false claims using evidence-aware deep learning. arXiv preprint arXiv:1809.06416 (2018)
39. Hovy, D.: The enemy in your own camp: how well can we detect statistically-generated fake reviews–an adversarial study. In: Proceedings of the 54th Annual Meeting of the Association for Computational Linguistics (Volume 2: Short Papers), pp. 351–356 (2016)
40. Solaiman, I., Brundage, M., Clark, J., Askell, A., Herbert-Voss, A., Wu, J., Radford, A., Wang, J.: Release strategies and the social impacts of language models. arXiv preprint arXiv:1908.09203 (2019)
41. Qian, F., Gong, C., Sharma, K., Liu, Y.: Neural user response generator: fake news detection with collective user intelligence. In: International Joint Conference on Artificial Intelligence (IJCAI), pp. 3834–3840 (2018)
42. Wang, Y., Yao, Q.: Few-shot learning: a survey. arXiv preprint arXiv:1904.05046 (2019)

43. Wang, Y., Ma, F., Jin, Z., Yuan, Y., Xun, G., Jha, K., Su, L., Gao, J.: EANN: event adversarial neural networks for multi-modal fake news detection. In: Proceedings of the 24th ACM SIGKDD International Conference on Knowledge Discovery & Data Mining, pp. 849–857. ACM (2018)
44. Hosseinimotlagh, S., Papalexakis, E.E.: Unsupervised content-based identification of fake news articles with tensor decomposition ensembles. MIS2, Marina Del Rey (2018)
45. Yang, S., Shu, K., Wang, S., Gu, R., Wu, F., Liu, H.: Unsupervised fake news detection on social media: a generative approach. In: AAAI (2019)

Part I
User Engagements in the Dissemination of Information Disorder

Discover Your Social Identity from What You Tweet: A Content Based Approach

Binxuan Huang and Kathleen M. Carley

Abstract An identity denotes the role an individual or a group plays in highly differentiated contemporary societies. In this paper, our goal is to classify Twitter users based on their role identities. We first collect a coarse-grained public figure dataset automatically, then manually label a more fine-grained identity dataset. We propose a hierarchical self-attention neural network for Twitter user role identity classification. Our experiments demonstrate that the proposed model significantly outperforms multiple baselines. We further propose a transfer learning scheme that improves our model's performance by a large margin. Such transfer learning also greatly reduces the need for a large amount of human labeled data.

Keywords Social identity · Twitter · User profiling · Text mining · Neural network

1 Introduction

An identity is a characterization of the role an individual takes on. It is often described as the social context specific personality of an individual actor or a group of people [2]. Identities can be things like jobs (e.g. "lawyer", "teacher"), gender (man, woman), or a distinguishing characteristic (e.g. "a shy boy", "a kind man"). People with different identities tend to exhibit different behaviors in the social space [10]. In this paper, we use role identity to refer to the roles individuals or groups play in society.

Specifically on social media platforms, there are many different kinds of actors using social media, e.g., people, organizations, and bots. Each type of actor has different motivations, different resources at their disposal, and may be under different internal policies or constraints on when they can use social media, how they

B. Huang (✉) · K. M. Carley
Carnegie Mellon University, Pittsburgh, PA, USA
e-mail: binxuanh@cs.cmu.edu; kathleen.carley@cs.cmu.edu

© Springer Nature Switzerland AG 2020   23
K. Shu et al. (eds.), *Disinformation, Misinformation, and Fake News in Social Media*,
Lecture Notes in Social Networks, https://doi.org/10.1007/978-3-030-42699-6_2

can represent themselves, and what they can communicate. If we want to understand who is controlling the conversation and whom is being impacted, it is important to know what types of actors are doing what.

To date, for Twitter, most research has separated types of actors largely based on whether the accounts are verified by Twitter or not [19], or whether they are bots or not [14]. Previous study has shown that separating Twitter users into bots and non-bots provides better understanding of U.S. presidential election online discussion [7]. Bessi and Ferrara reveal that social bots distort the 2016 U.S. presidential election online discussion and about one-fifth of the entire conversation comes from bots. However, a variety of different types of actors may be verified – e.g., news agencies, entertainment or sports team, celebrities, and politicians. Similarly, bots can vary – e.g., news bots and non-news bots. If we could classify the role identities of actors on Twitter, we could gain an improved understanding of who was doing the influencing and who was being influenced [12]. For example, knowing the social roles of bots would enable a more in-depth analysis of bot activities in the diffusion process of disinformation, eg. whether bots pretend to be news agencies to persuade regular users.

Understanding the sender's role is critical for doing research on, and developing technologies to stop, disinformation [11, 28]. Research has shown that disinformation has a greater reach if it is spread by news agencies and celebrities [33]. Disinformation is generally thought to be promoted by bots [3, 42]; however, most tools for identifying bots have relatively low accuracy when used in the wild [6]. News reporters, news agencies and celebrities often look like bots. Separating them out gives a better understanding of the role of bots in promoting disinformation. Assessing the extent to which official sites are communicating in a way that effectively counters disinformation also required identification of the sender's role. Thus, role identification is foundational for disinformation research

In this paper, the primary goal is to classify Twitter users based on their role identities on social media. First, we introduce two datasets for Twitter user identity classification. One is automatically collected from Twitter aiming at identifying public figures on social media. Another is a human labeled dataset for more fine-grained Twitter user identity classification, which includes identities like government officials, news reporters, etc. Second, we present a hierarchical self-attention neural network for Twitter user identity classification. In our experiments, we show our method achieves excellent results when compared to many strong classification baselines. Last but not least, we propose a transfer learning scheme for fine-grained user identity classification which boosts our model's performance a lot.

2   Related Work

Sociologists have long been interested in the usage of identities across various social contexts [40]. As summarized in [39], three relatively distinct usages of *identity* exist in the literature. Some use identity to refer to the culture of a people [9]. Some use it to refer to common identification with a social category [41]. While others use identity to refer to the role a person plays in highly differentiated contemporary societies. In this paper, we use the third meaning. Our goal for identity classification is to separate actors with different roles in online social media.

Identity is the way that individuals and collectives are distinguished in their relations with others [24]. Certain difficulties still exist for categorizing people into different groups based on their identities. Recasens et al. argue that identity should be considered to be varying in granularity and a categorical understanding would limit us in a fixed scope [35]. While much work could be done along this line, at this time we adopt a coarse-grained labeling procedure, that only looks at major identities in the social media space.

Twitter, a popular online news and social networking site, is also a site that affords interactive identity presentation to unknown audiences. As pointed out by Robinson et al., individuals form new cyber identities on the internet, which are not necessarily the way they would be perceived offline [36]. A customized identity classifier is needed for online social media like Twitter.

A lot of research has tried to categorize Twitter users based on certain criteria [34], like gender [8], location [22, 23, 44], occupation [21, 31], and political orientation [16]. Another similar research topic is bot detection [14], where the goal is to identify automated user accounts from normal Twitter accounts. Differing from them, our work tries to categorize Twitter users based on users' social identity or social roles. Similarly, Pirante et al. also study identity classification on Twitter [32]. However, their approach is purely based on profile description, while we combine user self-description and tweets together. Additionally, we demonstrate that tweets are more helpful for identity classification than personal descriptions in our experiments.

In fact, learning Twitter users' identities can benefit other related tasks. Twitter is a social media where individual user accounts and organization accounts co-exist. Many user classification methods may not work on these organization accounts, e.g., gender classification. Another example is bot detection. In reality, accounts of news agencies and celebrities often look like bots [15], because these accounts often employ automated services or teams (so called cyborgs), and they also share features with certain classes of bots; e.g., they may be followed more than they follow. Being able to classify actors' roles on Twitter would improve our ability to automatically differentiate pure bots from celebrity accounts.

3   Method

In this section, we describe details of our hierarchical self-attention neural networks. The overall architecture is shown in Fig. 1. Our model first maps each word into a low dimension word embedding space, then it uses a Bidirectional Long Short-Term Memory (Bi-LSTM) network [20] to extract context specific semantic representations for words. Using several layers of multi-head attention neural

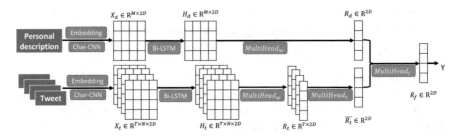

Fig. 1 The architecture of hierarchical self-attention neural networks for identity classification

networks, it generates a final classification feature vector. In the following parts, we elaborate these components in details.

3.1 Word Embedding

Our model first maps each word in description and tweets into a word embedding space $\in R^{V \times D}$ by a table lookup operation, where V is the vocabulary size, and D is the embedding dimension.

Because of the noisy nature of tweet text, we further use a character-level convolutional neural network to generate character-level word embeddings, which are helpful for dealing with out of vocabulary tokens. More specifically, for each character c_i in a word $w = (c_1, \ldots, c_k)$, we first map it into a character embedding space and get $v_{c_i} \in R^d$. Then a convolutional neural network is applied to generate features from characters [26]. For a character window $v_{c_i:c_{i+h-1}} \in R^{h \times d}$, a feature θ_i is generated by $\theta_i = f(w \cdot v_{c_i:c_{i+h-1}} + b)$ where $w \in R^{h \times d}$ and b are a convolution filter and a bias term respectively, $f(\cdot)$ is a non-linear function $relu$. Sliding the filter from the beginning of the character embedding matrix till the end, we get a feature vector $\theta = [\theta_1, \theta_2, \ldots, \theta_{k-h+1}]$. Then, we apply max pooling over this vector to get the most representative feature. With D such convolutional filters, we get the character-level word embedding for word w.

The final vector representation $v_w \in R^{2D}$ for word w is just the concatenation of its general word embedding vector and character-level word embedding vector. Given one description with M tokens and T tweets each with N tokens, we get two embedding matrices $X_d \in R^{M \times 2D}$ and $X_t \in R^{T \times N \times 2D}$ for description and tweets respectively.

3.2 Bi-LSTM

After get the embedding matrices for tweets and description, we use a bidirectional LSTM to extract context specific features from each text. At each time step, one forward LSTM takes the current word vector v_{w_i} and the previous hidden state $\overrightarrow{h_{w_{i-1}}}$ to generate the hidden state for word w_i. Another backward LSTM generates another sequence of hidden states in the reversed direction. We also tried Bi-directional GRU [13] in our initial experiments, which yields slightly worse performance.

$$\overrightarrow{h_{w_i}} = \overrightarrow{LSTM}(v_{w_i}, \overrightarrow{h_{w_{i-1}}})$$
$$\overleftarrow{h_{w_i}} = \overleftarrow{LSTM}(v_{w_i}, \overleftarrow{h_{w_{i+1}}})$$
(1)

The final hidden state $h_{w_i} \in R^{2D}$ for word w_i is the concatenation of $\overrightarrow{h_{w_i}}$ and $\overleftarrow{h_{w_i}}$ as $h_{w_i} = [\overrightarrow{h_{w_i}}, \overleftarrow{h_{w_i}}]$. With T tweets and one description, we get two hidden state matrices $H_t \in R^{T \times N \times 2D}$ and $H_d \in R^{M \times 2D}$.

3.3 Attention

Following the Bi-LSTM layer, we use a word-level multi-head attention layer to find important words in a text [43].

Specifically, a multi-head attention is computed as follows:

$$MultiHead(H_d) = Concat(head_1, \ldots, head_h)W^O$$

$$head_i = softmax(\frac{H_d W_i^Q \cdot (H_d W_i^K)^T}{\sqrt{d_k}})H_d W_i^V$$

where $d_k = 2D/h$, W_i^Q, W_i^K, $W_i^V \in R^{2D \times d_k}$, and $W^O \in R^{hd_k \times 2D}$ are projection parameters for query, key, value, and output respectively.

Take a user description for example. Given the hidden state matrix H_d of the description, each head first projects H_d into three subspaces – query $H_d W_i^Q$, key $H_d W_i^K$, and value $H_d W_i^V$. The matrix product between key and query after softmax normalization is the self-attention, which indicates important parts in the value matrix. The multiplication of self-attention and value matrix is the output of this attention head. The final output of multi-head attention is the concatenation of h such heads after projection by W^O.

After this word-level attention layer, we apply a row-wise average pooling to get a high-level representation vector for description.

$$R_d = row\_avg(MultiHead_w(H_d)) \in R^{2D}$$
(2)

Similarly, we can get T representation vectors from T tweets using the same word-level attention, which forms $R_t \in R^{T \times 2D}$.

Further, a tweet-level multi-head attention layer computes the final tweets representation vector $\bar{R}_t$ as follows:

$$\bar{R}_t = row\_avg(MultiHead_t(R_t)) \in R^{2D} \tag{3}$$

In practise, we also tried using an additional Bi-LSTM layer to model the sequence of tweets, but we did not observe any significant performance gain.

Given the description representation R_d and tweets representation $\bar{R}_t$, a field attention generates the final classification feature vector

$$R_f = row\_avg(MultiHead_f([R_d; \bar{R}_t])) \tag{4}$$

where $[R_d; \bar{R}_t] \in R^{2 \times 2D}$ means concatenating by row.

3.4 Final Classification

Finally, the probability for each identity is computed by a softmax function:

$$P = softmax(WR_f + b) \tag{5}$$

where $W \in R^{|C| \times 2D}$ is the projection parameter, $b \in R^{|C|}$ is the bias term, and C is the set of identity classes. We minimize the cross-entropy loss function to train our model,

$$loss = -\sum_{c \in C} Y_c \cdot \log P_c \tag{6}$$

where Y_c equals to 1 if the identity is of class c, otherwise 0.

4 Experiments

4.1 Dataset

To examine the effectiveness of our method, we collect two datasets from Twitter. The first is a public figure dataset. We use Twitter's verification as a proxy for public figures. These verified accounts include users in music, government, sports, business, and etc.[1] We sampled 156746 verified accounts and 376371 unverified

---

[1]https://help.twitter.com/en/managing-your-account/about-twitter-verified-accounts

Table 1 A brief summary of our two datasets

	Public figure		Identity						
	Verified	Unverified	Media	Reporter	Celebrity	Government	Company	Sport	Regular
Train	152,368	365,749	1140	614	876	844	879	870	6623
Dev.	1452	3548	52	23	38	40	35	43	269
Test	2926	7074	97	39	75	81	66	74	568

Table 2 Five representative Twitter handles for each identity class except for regular users

News media	News reporter	Celebrity	Government official	Company	Sport
CBSNews	PamelaPaulNYT	aliciakeys	USDOL	VisualStudio	NBA
earthtimes	HowardKurtz	Adele	RepRichmond	lifeatgoogle	Pirates
BBCNewsAsia	jennaportnoy	GreenDay	HouseGOP	BMW	NFL
phillydotcom	wpjenna	ladygaga	BelgiumNATO	AEO	KKRiders
TheSiasatDaily	twithersAP	TheEllenShow	usafpressdesk	Sony	USAGym

accounts through Twitter's sample stream data.[2] Then we collected their most recent 20 tweets from Twitter's API in November 2018. We randomly choose 5000 users as a development set and 10,000 users as a test set. A summary of this dataset is shown in Table 1.

In addition, we introduce another human labeled identity dataset for more fine-grained identity classification, which contains seven identity classes: "news media", "news reporter", "government official", "celebrity", "company", "sport", and "regular user". For each identity, we manually labelled thousands of Twitter users and collected their most recent 20 tweets for classification in November 2018. For the regular Twitter users, we randomly sampled them from the Twitter sample stream. News media accounts are these official accounts of news websites like BBC. News reporters are mainly composed of news editors or journalists. Government officials represent government offices or politicians. We collected these three types of accounts from corresponding official websites. For the other three categories, we first search Twitter for these three categories, and then we downloaded their most recent tweets using Twitter's API. Two individual workers labeled these users independently, and we include users that both two workers agreed on. The inter-rater agreement measure is 0.96. In Table 2, we list several representative Twitter handles for each identity class except for regular users. Table 1 shows a summary of this dataset. We randomly select 500 and 1000 users for development and test respectively. Since regular users are the majority of Twitter users, about half of the users in this dataset are regular users.

This paper focuses on a content-based approach for identity classification, so we only use personal description and text of each tweet for each user.

---

[2]https://developer.twitter.com/en/docs/tweets/sample-realtime/overview/GET_statuse_sample.html

4.2   Hyperparameter Setting

In our experiments, we initialize the general word embeddings with released 300-dimensional Glove vectors[3] [29]. For words not appearing in Glove vocabulary, we randomly initialize them from a uniform distribution $U(-0.25, 0.25)$. The 100-dimensional character embeddings are initialized with a uniform distribution $U(-1.0, 1.0)$. These embeddings are adapted during training. We use filter windows of size 3,4,5 with 100 feature maps each. The state dimension D of LSTM is chosen as 300. For all the multi-head attention layers, we choose the number of heads as 6. We apply dropout [38] on the input of Bi-LSTM layer and also the output of the softmax function in these attention layers. The dropout rate is chosen as 0.5. The batch size is 32. We use Adam update rule [27] to optimize our model. The initial learning rate is 10^{-4} and it drops to 10^{-5} at the last 1/3 epochs. We train our model 10 epochs, and every 100 steps we evaluate our method on development set and save the model with the best result. All these hyperparameters are tuned on the development set of identity dataset. We implemented our model using Tensorflow [1] on a Linux machine with Titan XP GPUs.

4.3   Baselines

MNB: Multinomial Naive Bayes classifier with unigrams and bigrams. The term features are weighted by their TF-IDF scores. Additive smoothing parameter is set as 10^{-4} via a grid search on the development set of identity dataset.

SVM: Support Vector Machine classifier with unigrams and linear kernel. The term features are weighted by their TF-IDF scores. Penalty parameter is set as 100 via a grid search on the development set of identity dataset.

CNN: Convolutional Neural Networks [26] with filter window size 3,4,5 and 100 feature maps each. Initial learning rate is 10^{-3} and drops to 10^{-4} at the last 1/3 epochs.

Bi-LSTM: Bidirectional-LSTM model with 300 hidden states in each direction. The average of output at each step is used for the final classification.

Bi-LSTM-ATT: Bidirectional-LSTM model enhanced with self-attention. We use multi-head attention with 6 heads.

fastText [25]: we set word embedding size as 300, use unigram, and train it 10 epochs with initial learning 1.0.

For methods above, we combine personal description and tweets into a whole document for each user.

---

[3]https://nlp.stanford.edu/projects/glove/

Table 3 Comparisons between our methods and baselines

		Public figure		Identity	
		Accuracy	Macro-F1	Accuracy	Macro-F1
Baselines	MNB	81.81	82.79	82.9	75.91
	SVM	90.60	88.59	85.9	80.19
	fastText	90.93	89.01	85.7	80.01
	CNN	91.45	89.85	85.9	81.24
	Bi-LSTM	93.10	91.84	86.5	84.25
	Bi-LSTM-ATT	93.23	91.94	87.3	83.35
Ablated models	w/o attentions	93.78	92.45	87.0	83.26
	w/o charcnn	93.47	92.23	89.0	85.39
	w/o description	92.39	90.90	86.7	81.56
	w/o tweets	91.62	89.77	84.2	78.41
	Full model	**94.21**	**93.07**	**89.5**	**86.09**
	Full model-transfer			**91.6**	**88.63**

4.4 Results

In Table 3, we show comparison results between our model and baselines. Generally, LSTM based methods work the best among all these baseline approaches. SVM has comparable performance to these neural network based methods on the identity dataset, but falls behind on the larger public figure dataset.

Our method outperforms these baselines on both datasets, especially for the more challenging fine-grained identity classification task. Our model can successfully identify public figures with accuracy 94.21% and classify identity with accuracy 89.5%. Compared to a strong baseline Bi-LSTM-ATT, our model achieves a 2.2% increase in accuracy, which shows that our model with structured input has better classification capability.

We further performed ablation studies to analyze the contribution of each model component, where we removed attention modules, character-level word embeddings, tweet texts, and user description one by one at a time. As shown in Table 3, attention modules make a great contribution to the final classification performance, especially for the more fine-grained task. We present the performance breakdown for each attention module in Table 4. Each level of attention effectively improves the performance of our model. Recognizing important words, tweets, and feature fields at different levels is helpful for learning classification representations. According to Table 3, the character-level convolutional layer is also helpful for capturing some character-level patterns.

We also examined the impact of two different text fields: personal description and tweets. Indeed, we found that what users tweeted about is more important than what they described themselves. On both datasets, users' tweets provide more discriminative power than users' personal descriptions.

Table 4 The effectiveness of different levels of attentions tested on the identity dataset

	Accuracy	Macro-F1
Full Model	89.5	86.09
w/o word attention	88.8	84.41
w/o field attention	88.5	85.24
w/o tweet attention	88.5	84.6
w/o all attention	87.0	83.26

4.5 Transfer Learning for Fine-Grained Identity Classification

In reality, it is expensive to get a large-scale human labeled dataset for training a fine-grained identity classifier. However, a well-known drawback of neural network based methods is that they require a lot of data for training. Recently, learning from massive data and transferring learned knowledge to other tasks attracts a lot of attention [17, 30]. Since it is relatively easier to get a coarse-grained identity dataset to classify those public figures, we explore how to use this coarse-grained public figure dataset to help the training of fine-grained identity classifier.

Specifically, we first pretrain a binary classifier on the public figure dataset and save the best trained model on its development set. To make a fair comparison, we excluded all the users appearing in identity dataset from the public figure dataset when we built our datasets. Then we initialize the parameters of fine-grained identity classifier with this pretrained model except for the final classification layer. After such initialization step, we first train the final classification layer for 3 epochs with learning rate 0.01, and then train our full identity classification model with the same procedure as before. We observe a big performance boost when we apply such pretraining as shown in Table 3. The classification accuracy for the fine-grained task increases by 2.1% with transfer learning.

We further examined the performance of our model with pretraining using various amounts of training data. As shown in Fig. 2, our pretrained model reaches a comparable performance only with 20–30% labeled training data when compared to the model trained on full identity dataset without pretraining. Using only 20% of training data, we can get accuracy 0.888 and F1 0.839. If we increase the data size to 30% of the training data, the accuracy and F1 will increase to 0.905 and 0.863 respectively. Such pretraining makes great improvements over fine-grained identity classification especially when we lack labeled training data.

4.6 Case Study

In this section, we present a case study in the test set of identity dataset to show the effectiveness of our model. Because of the difficulties of visualizing and interpreting multi-head attention weights, we instead average over the attention weights in multiple heads which gives us an approximation of the importance of each word in

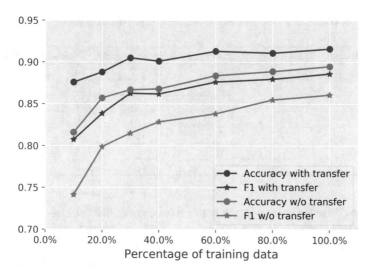

Fig. 2 Performance comparison between our model with transfer learning and without. We train our model on various amounts of training data

Fig. 3 The visualization of attention weights for each tweet and description. The color depth denotes the importance degree of a word per tweet. The importance of each tweet is depicted as the background color of corresponding tweet header

texts. Take the user description for example, the approximated importance weight of each word in the description is given by

$$\alpha_d = row\_avg(\frac{1}{h} \sum_i softmax(\frac{H_d W_i^Q (H_d W_i^K)^T}{\sqrt{d_k}}))$$

Similarly, we can get the importance weights for tweets as well as words in tweets.

In Fig. 3, we show twenty tweets and a description from a government official user. We use the background color to represent importance weight for each word. The color depth denotes the importance degree of a word per tweet. We plot the tweet-level importance weights as the background color of tweet index at the

Table 5 The confusion matrix generated by our best trained model from the test set of identity dataset

TruthPrediction	Regular	Media	Celebrity	Sport	Company	Government	Reporter
Regular	535	10	12	0	5	2	4
Media	6	81	1	4	2	2	1
Celebrity	15	0	55	2	2	1	0
Sport	1	1	0	71	1	0	0
Company	1	2	1	4	58	0	0
Government	1	1	0	0	0	79	0
Reporter	1	0	1	0	0	0	37

beginning of each tweet. As shown in this figure, words like "congressman", "legislation" in this user's description are important clues indicating his/her identity. From the tweet-level attention, we know that 8th and 14th tweets are the most important tweets related with the identity because they include words like "legislation" and "bipartisan". On the contrary, 5th tweet of this user only contain some general words like "car", which makes it less important than other tweets.

4.7 Error Analysis

We perform an error analysis to investigate why our model fails in certain cases. Table 5 shows the confusion matrix generated from prediction results of our identity dataset. As shown in this table, it is relatively harder for our model to distinguish between celebrities and regular users. We further looked at such errors with high confidences and found that some celebrities just have not posted any indicating words in their tweets or descriptions. For example, one celebrity account only use "A Virgo" in the description without any other words, which makes this account predicted as a regular user. Including other features like number of followers or network connections may overcome this issue, and we leave it for future work. Another common error happens when dealing with non-English tweets. Even enhanced with transferred knowledge from the large-scale verify dataset, our model still cannot handle some rare languages in the data.

5 Discussion and Conclusion

As previously discussed, identities can vary in granularity. We examined two levels – coarse grained (verified or not) and more fine grained (news media, government officials, etc.). However, there could be more levels. This limits our understanding of activities of online actors with those identities. A hierarchical approach for identity classification might be worth further research. Future research should take this into consideration and learn users' identities in a more flexible way. Besides, because of the nature of social media, the content on Twitter would evolve rapidly. In order to deploy our method in real-time, we need consider an online learning procedure that adapts our model to new data patterns. Since our method is purely content-based, potential improvements could be made using additional information like the number of users' followers, users' network connections, and even their profile images. We leave this as our future work.

In the real-world people often have multiple identities – e.g., Serbian, Entrepreneur, Policewoman, Woman, Mother. The question is what is the relation between identities, users, and user accounts. Herein, we treat each account as a different user. However, in social media, some people use different accounts and/or different social media platforms for different identities – e.g., Facebook for Mother, Twitter for Entrepreneur and a separate Twitter handle for official policewoman account. In this paper, we made no effort to determine whether an individual had multiple accounts. Thus, the same user may get multiple classifications if that user has multiple accounts. Future work should explore how to link multiple identities to the same user. To this point, when there is either a hierarchy of identities or orthogonal identity categories, then using identities at different levels of granularity, as we did herein, enables multiple identities to be assigned to the same account and so to the same user.

In conclusion, we introduce two datasets for online user identity classification. One is automatically extracted from Twitter, the other is a manually labelled dataset. We present a novel content-based method for classifying social media users into a set of identities (social roles) on Twitter. Our experiments on two datasets show that our model significantly outperforms multiple baseline approaches. Using one personal description and up to twenty tweets for each user, we can identify public figures with accuracy 94.21% and classify more fine-grained identities with accuracy 89.5%. We proposed and tested a transfer learning scheme that further boosts the final identity classification accuracy by a large margin. Though, the focus of this paper is learning users' social identities. It is possible to extend this work to predict other demographics like gender and age.

References

1. Abadi, M., Barham, P., Chen, J., Chen, Z., Davis, A., Dean, J., Devin, M., Ghemawat, S., Irving, G., Isard, M., et al.: Tensorflow: a system for large-scale machine learning. In: 12th USENIX Symposium on Operating Systems Design and Implementation (OSDI 16), pp. 265–283 (2016)
2. Ashforth, B.E., Mael, F.: Social identity theory and the organization. Acad. Manage. Rev. 14(1), 20–39 (1989)
3. Babcock, M., Beskow, D.M., Carley, K.M.: Beaten up on twitter? exploring fake news and satirical responses during the black panther movie event. In: International Conference on Social Computing, Behavioral-Cultural Modeling and Prediction and Behavior Representation in Modeling and Simulation, pp. 97–103. Springer (2018)

4. Benigni, M., Joseph, K., Carley, K.M.: Mining online communities to inform strategic messaging: practical methods to identify community-level insights. Comput. Math. Organ. Theory **24**(2), 224–242 (2018)
5. Benigni, M.C., Joseph, K., Carley, K.M.: Bot-ivistm: assessing information manipulation in social media using network analytics. In: Emerging Research Challenges and Opportunities in Computational Social Network Analysis and Mining, pp. 19–42. Springer, Cham (2019)
6. Beskow, D.M., Carley, K.M.: Bot conversations are different: leveraging network metrics for bot detection in twitter. In: 2018 IEEE/ACM International Conference on Advances in Social Networks Analysis and Mining (ASONAM), pp. 825–832. IEEE (2018)
7. Bessi, A., Ferrara, E.: Social bots distort the 2016 us presidential election online discussion. First Monday **21**(11–17) (2016)
8. Burger, J.D., Henderson, J., Kim, G., Zarrella, G.: Discriminating gender on twitter. In: Proceedings of the Conference on Empirical Methods in Natural Language Processing, pp. 1301–1309. Association for Computational Linguistics (2011)
9. Calhoun, C.J.: Social Theory and the Politics of Identity. Wiley-Blackwell, Oxford (1994)
10. Callero, P.L.: Role-identity salience. Soc. Psychol. Q. **48**(3), 203–215 (1985)
11. Carley, K.M., Cervone, G., Agarwal, N., Liu, H.: Social cyber-security. In: International Conference on Social Computing, Behavioral-Cultural Modeling and Prediction and Behavior Representation in Modeling and Simulation, pp. 389–394. Springer (2018)
12. Cha, M., Haddadi, H., Benevenuto, F., Gummadi, P.K., et al.: Measuring user influence in twitter: the million follower fallacy. Icwsm **10**(10–17), 30 (2010)
13. Cho, K., Van Merriënboer, B., Gulcehre, C., Bahdanau, D., Bougares, F., Schwenk, H., Bengio, Y.: Learning phrase representations using RNN encoder-decoder for statistical machine translation. arXiv preprint arXiv:1406.1078 (2014)
14. Chu, Z., Gianvecchio, S., Wang, H., Jajodia, S.: Who is tweeting on twitter: human, bot, or cyborg? In: Proceedings of the 26th Annual Computer Security Applications Conference, pp. 21–30. ACM (2010)
15. Chu, Z., Gianvecchio, S., Wang, H., Jajodia, S.: Detecting automation of twitter accounts: are you a human, bot, or cyborg? IEEE Trans. Dependable Secure Comput. **9**(6), 811–824 (2012)
16. Colleoni, E., Rozza, A., Arvidsson, A.: Echo chamber or public sphere? predicting political orientation and measuring political homophily in twitter using big data. J. Commun. **64**(2), 317–332 (2014)
17. Devlin, J., Chang, M.W., Lee, K., Toutanova, K.: Bert: pre-training of deep bidirectional transformers for language understanding. arXiv preprint arXiv:1810.04805 (2018)
18. Heise, D., MacKinnon, N.: Self, Identity, and Social Institutions. Palgrave Macmillan, New York (2010)
19. Hentschel, M., Alonso, O., Counts, S., Kandylas, V.: Finding users we trust: scaling up verified twitter users using their communication patterns. In: ICWSM (2014)
20. Hochreiter, S., Schmidhuber, J.: Long short-term memory. Neural Comput. **9**(8), 1735–1780 (1997)
21. Hu, T., Xiao, H., Luo, J., Nguyen, T.V.T.: What the language you tweet says about your occupation. In: Tenth International AAAI Conference on Web and Social Media (2016)
22. Huang, B., Carley, K.M.: On predicting geolocation of tweets using convolutional neural networks. In: International Conference on Social Computing, Behavioral-Cultural Modeling and Prediction and Behavior Representation in Modeling and Simulation, pp. 281–291. Springer (2017)
23. Huang, B., Carley, K.: A hierarchical location prediction neural network for twitter user geolocation. In: Proceedings of the 2019 Conference on Empirical Methods in Natural Language Processing and the 9th International Joint Conference on Natural Language Processing (EMNLP-IJCNLP), pp. 4731–4741. Association for Computational Linguistics (2019). https://doi.org/10.18653/v1/D19-1480
24. Jenkins, R.: Social identity. Routledge, London (2014)
25. Joulin, A., Grave, E., Bojanowski, P., Mikolov, T.: Bag of tricks for efficient text classification. arXiv preprint arXiv:1607.01759 (2016)

26. Kim, Y.: Convolutional neural networks for sentence classification. arXiv preprint arXiv:1408.5882 (2014)
27. Kingma, D.P., Ba, J.: Adam: a method for stochastic optimization. arXiv preprint arXiv:1412.6980 (2014)
28. National Academies of Sciences, Engineering, and Medicine (U.S.).: A Decadal Survey of the Social and Behavioral Sciences: A Research Agenda for Advancing Intelligence Analysis. The National Academies Press, Washington (2019). https://doi.org/10.17226/25335
29. Pennington, J., Socher, R., Manning, C.: Glove: global vectors for word representation. In: Proceedings of the 2014 Conference on Empirical Methods in Natural Language Processing (EMNLP), pp. 1532–1543 (2014)
30. Peters, M.E., Neumann, M., Iyyer, M., Gardner, M., Clark, C., Lee, K., Zettlemoyer, L.: Deep contextualized word representations. arXiv preprint arXiv:1802.05365 (2018)
31. Preoţiuc-Pietro, D., Lampos, V., Aletras, N.: An analysis of the user occupational class through twitter content. In: Proceedings of the 53rd Annual Meeting of the Association for Computational Linguistics and the 7th International Joint Conference on Natural Language Processing (Volume 1: Long Papers), vol. 1, pp. 1754–1764 (2015)
32. Priante, A., Hiemstra, D., van den Broek, T., Saeed, A., Ehrenhard, M., Need, A.: # whoami in 160 characters? classifying social identities based on twitter profile descriptions. In: Proceedings of the First Workshop on NLP and Computational Social Science, pp. 55–65 (2016)
33. Ramon Villa Cox, M.B., Carley, K.M.: Pretending positive, pushing false: comparing captain marvel misinformation campaigns. Fake News, Disinformation, and Misinformation in Social Media-Emerging Research Challenges and Opportunities (2019)
34. Rangel Pardo, F.M., Celli, F., Rosso, P., Potthast, M., Stein, B., Daelemans, W.: Overview of the 3rd author profiling task at pan 2015. In: CLEF 2015 Evaluation Labs and Workshop Working Notes Papers, pp. 1–8 (2015)
35. Recasens, M., Hovy, E., Martí, M.A.: Identity, non-identity, and near-identity: addressing the complexity of coreference. Lingua 121(6), 1138–1152 (2011)
36. Robinson, L.: The cyberself: the self-ing project goes online, symbolic interaction in the digital age. New Media Soc. 9(1), 93–110 (2007)
37. Smith-Lovin, L.: The strength of weak identities: social structural sources of self, situation and emotional experience. Soc. Psychol. Q. 70(2), 106–124 (2007)
38. Srivastava, N., Hinton, G., Krizhevsky, A., Sutskever, I., Salakhutdinov, R.: Dropout: a simple way to prevent neural networks from overfitting. J. Mach. Learn. Res. 15(1), 1929–1958 (2014)
39. Stryker, S., Burke, P.J.: The past, present, and future of an identity theory. Soc. Psychol. Q. 63(4), 284–297 (2000)
40. Tajfel, H.: Social identity and intergroup behaviour. Information (International Social Science Council) 13(2), 65–93 (1974)
41. Tajfel, H.: Social Identity and Intergroup Relations. Cambridge University Press, Cambridge (1982)
42. Uyheng, J., Carley, K.M.: Characterizing bot networks on twitter: an empirical analysis of contentious issues in the asia-pacific. In: International Conference on Social Computing, Behavioral-Cultural Modeling and Prediction and Behavior Representation in Modeling and Simulation, pp. 153–162. Springer (2019)
43. Vaswani, A., Shazeer, N., Parmar, N., Uszkoreit, J., Jones, L., Gomez, A.N., Kaiser, Ł., Polosukhin, I.: Attention is all you need. In: Advances in Neural Information Processing Systems, pp. 5998–6008 (2017)
44. Zhang, Y., Wei, W., Huang, B., Carley, K.M., Zhang, Y.: Rate: overcoming noise and sparsity of textual features in real-time location estimation. In: Proceedings of the 2017 ACM on Conference on Information and Knowledge Management, pp. 2423–2426. ACM (2017)

User Engagement with Digital Deception

Maria Glenski, Svitlana Volkova, and Srijan Kumar

Abstract Digital deception in online social networks, particularly the viral spread of misinformation and disinformation, is a critical concern at present. Online social networks are used as a means to spread digital deception within local, national, and global communities which has led to a renewed focus on the means of detection and defense. The audience (i.e., social media users) form the first line of defense in this process and it is of utmost importance to understand the who, how, and what of audience engagement. This will shed light on how to effectively use this wisdom-of-the-audience to provide an initial defense. In this chapter, we present the key findings of the recent studies in this area to explore user engagement with trustworthy information, misinformation, and disinformation framed around three key research questions (1) Who engages with mis- and dis-information?, (2) How quickly does the audience engage with mis- and dis-information?, and (3) What feedback do users provide? These patterns and insights can be leveraged to develop better strategies to improve media literacy and informed engagement with crowd-sourced information like social news.

Keywords Disinformation · Misinformation · User engagement

Social media platforms have gone beyond means of entertainment or social networking to become commonly used mechanisms for social news consumption. As the reliability or trustworthiness of media, news organizations, and other sources is increasingly debated, the society's reliance on social media as a primary source for news, opinion, and information has triggered renewed attention on the spread of misinformation. In particular, in these online communities with increased

M. Glenski · S. Volkova
Data and Analytics Group, National Security Directorate, Pacific Northwest National Laboratory, Richland, WA, USA
e-mail: maria.glenski@pnnl.gov; svitlana.volkova@pnnl.gov

S. Kumar (✉)
Georgia Institute of Technology, Atlanta, GA, USA

© Springer Nature Switzerland AG 2020
K. Shu et al. (eds.), *Disinformation, Misinformation, and Fake News in Social Media*, Lecture Notes in Social Networks, https://doi.org/10.1007/978-3-030-42699-6_3

importance as a means of convenient and swift but potentially unreliable information acquisition – 68% of Americans report that they get at least some of their news from social media, however 57% of social media users who consume news on one or more of these platforms expect that the news they see to be "largely inaccurate" [28]. Most studies that investigate misinformation spread in social media focus on individual events and the role of the network structure in the spread [24, 25, 31, 47] or detection of false information [33]. Many related studies have focused on the language of misinformation in social media [18, 29, 32, 34, 43, 45] to detect types of deceptive news, compare the behavior of traditional and alternative media [37], or detect rumor-spreading users [33].

These studies have found that the size and shape of (mis) information cascades within a social network is heavily dependent on the initial reactions of the audience. Fewer studies have focused on understanding how users react to news sources of varying credibility and how their various response types contribute to the spread of (mis)information. An obvious, albeit challenging, way to do so is through the audience and their complex social behavior—the individuals and society as a whole who consume, disseminate, and act on the information they receive. However, there are a few challenges that complicate the task of using the audience as reliable signals of misinformation detection.

The first major challenge is that users have a *truth bias* wherein they tend to believe that others are telling the truth [14, 39]. Furthermore, research has found that being presented with brief snippets or clips of information (the style of information most commonly consumed on social media) exacerbates this bias [39]. Although this bias is reduced as individuals make successive judgements about veracity [39], social media users tend to make shallow, single engagements with content on social media [5, 6]. In fact, recent studies have found that 59% of bitly-URLs on Twitter are shared without ever being read [5] and 73% of Reddit posts were voted on without reading the linked article [6].

The second major challenge is that humans are not perfect in identifying false information when they come across it while browsing. Recent study by Kumar et al. [22, 23] showed that when people are in the reading mode, they can effectively detect false information. In particular, an experiment done with Wikipedia hoaxes showed that humans achieved 66% accuracy in distinguishing hoaxes from non-hoaxes, compared to 50% accuracy by random guessing. While they are better than random, humans make a mistake once out of every three attempts to detect false information, which can add error to the crowd-sourced human intelligence. The real strength lies in signals from many consumers at the same time, instead of a single individual.

The third major challenge is that users attempt to counterbalance their shallow engagement with *content* with a reliance on the crowd-provided commentary for information about the content and its credibility. When users do so, they rely on the assumption that these social media platforms are able to leverage the *wisdom of the crowd* to crowd-source reliable ratings, rankings, or other curation of information so that users don't need to expend the cognitive resources to do so themselves for the deluge of information flooding their subreddits, timelines, or news feeds. However, other research has illustrated how the *hive mind* or *herd mentality* observed when individuals' perceptions of quality or value follow the behavior of a group can be suboptimal for the group and individual members alike [1, 15, 27]. Studies have also found that user behavior (and thus, the content that is then shown to other users) can be easily influenced and manipulated through injections of artificial ratings [7, 8, 30, 46].

The audience (i.e., social media users) is effectively the "first line of defense" against the negative impacts and spread of misinformation or digital deception. It is important not only to understand how disinformation spreads or gains rapid traction (i.e., "goes viral") and how to identify digital deception in a variety of forms but also how individuals and the audience in general currently react, engage, and amplify the reach of deception. These patterns and insights can be leveraged to better develop strategies to improve media literacy and informed engagement with crowd-sourced information like social news. In this chapter, we highlight several recent studies that focus on the human element (the audience) of the (mis) and (dis)information ecosystem and news cycle.

As reliance on social media as a source of news remains consistently high and the reliability of news sources is increasingly debated, it is important to understand not only what (mis) and (dis)information is produced, how to identify digital deception at coarse and fine granularities, and which algorithmic or network characteristics enable its spread, but also how users (human and automated alike) consume and contribute to the (mis) and (dis)information cycle. For example, how do users react to news sources of varied levels of credibility and what commentary or kinds of reactions are presented to other users?

In this chapter, we highlight key findings from the studies summarized in Table 1 framed around three key research questions:

RQ1: *Who* engages with (mis) and (dis)information?,
RQ2: *What* kind of feedback do users provide?, and
RQ3: *How quickly* do users engage with (mis) and (dis)information?

in Sects. 2, 3, and 4, respectively. Before we explore these research questions, we first present an overview of the *Methods and Materials* used in Sect. 1.

Table 1 Studies highlighted in this chapter and the sections that reference each study

Reference	Title	Sections
[11]	Propagation from deceptive news sources: who shares, how much, how evenly, and how quickly?	2, 4
[10]	Identifying and understanding user reactions to deceptive and trusted social news sources	3, 4
[9]	How humans versus bots react to deceptive and trusted news sources: A case study of active users	2, 3, 4

1   Methods and Materials

As we noted above, most studies that examine digital deception spread focus on individual events such as natural disasters [40], political elections [4], or crises events [38] and examine the response to the event on specific social platforms. In contrast, the studies highlighted in this chapter consider users' engagement patterns across news sources identified as spreading trustworthy information versus disinformation – highlighting distinctions in audience composition or engagement patterns that can be leveraged for robust defense against (mis) and (dis)information, educational strategies to mitigate the continued spread or negative impacts of digital deception, and more. Before we highlight key findings, we present an overview of the processes used in the studies that will be referenced in the following sections.

1.1   Attributing News Sources

In several of the studies highlighted in the following section [9–11], credibility annotations partition news sources into (1) fine-grained or (2) coarse labeled sets based on the hierarchy of types of information spread in Fig. 1. Fine-grained labeled news sources are partitioned into five classes of news media. That is, news sources identified as a:

- *trustworthy* news source that provided factual information with no intent to deceive;

or one of several classes of *deceptive* news sources:

- *clickbait*: attention-grabbing, misleading, or vague headlines to attract an audience;

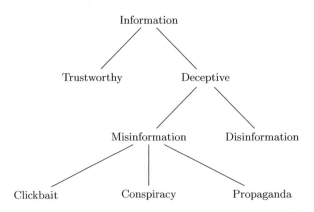

Fig. 1 Hierarchy of information, misinformation, and disinformation used in news source annotations

- *conspiracy theories*: uncorroborated or unreliable information to explain events or circumstances;
- *propaganda*: intentionally misleading information to advance a social or political agenda; or
- *disinformation*: fabricated and factually incorrect information spread with an intention to deceive the audience.

Coarse-grained labeled sets build off of these fine-grained annotations to look at more abstract groupings of:

- *trustworthy* news sources that provide factual information with no intent to deceive;
- *misinformation* news sources identified as spreading clickbait, conspiracy theories, and propaganda; or
- *misinformation* + *disinformation* news sources identified as spreading clickbait, conspiracy theories, propaganda, *and intentional disinformation*.

New sources identified as spreading disinformation were collected from EUvs-Disinfo.eu[1] while all others were obtained from a list compiled by Volkova et al. [43] through a combination of crowd-sourcing and public resources.[2] As of November 2016, EUvsDisinfo reports included almost 1,992 confirmed disinformation campaigns found in news reports from around Europe and beyond.

1.2 Inferring User Account Types: Automated Versus Manual

Classification of user accounts as manually run by individuals (i.e. human) or account that is automated (i.e. bot) is done through thresholding of botometer scores. Botometer scores [2] indicate the likelihood of a user account being an automated bot account and are collected for a given user. The label of 'bot' is then assigned if the score is at or above the bot threshold of 0.5, otherwise the label of 'human' is assigned to the user.

1.3 Predicting User Demographics

To infer gender, age, income, and education demographics of users identified to be individual, manually-run accounts, Glenski et al. [11] employed a neural network model trained on a large, previously annotated Twitter dataset [42]. Following

---

[1]News sources collected from EUvsDisinfor.eu were identified as spreaders of *disinformation* by the European Union's East Strategic Communications Task Force.

[2]Example resources used by Volkova et al [43] to compile deceptive news sources: http://www.fakenewswatch.com/, http://www.propornot.com/p/the-list.html.

previous methodology [42], each demographic attribute was assigned one of two mutually exclusive classes. Gender was classified as either male (M) or female (F), age as either younger than 25 (Y) or 25 and older (O), income as below (B) or at and above (A) $35,000 a year, and education as having only a high school education (H) or at least some college education (C).[3]

1.4 Measuring Inequality of User Engagement

In order to measure the inequality of engagement with trustworthy information versus deceptive news, we leverage three measures commonly used to measure income inequality: Lorenz curves, Gini coefficients, and Palma ratios. Rather than measuring how shares of a region, nation, or other population's income is spread across the individuals within the population, these metrics can be adapted to quantify and illustrate how interactions or the volume of engagement is spread across the population of users who engage with (mis) and (dis)information. This allows us to compare inequality of engagement with news sources across types of information (trustworthy news, conspiracy, disinformation etc.) in a approach to the way economists compare income inequality across countries.

Lorenz curves (an example of which is illustrated in Fig. 2) are often used as a graphical representation of income or wealth distributions [17]. In those domains, the curves plot the cumulative percentage of wealth, income, or some other variable

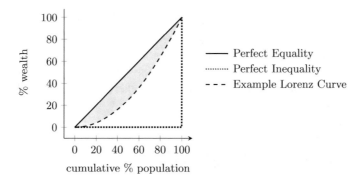

Fig. 2 Lorenz curves illustrate inequality within a frequency distribution graphically by plotting the cumulative share (from least to greatest) of the variable under comparison (e.g. income or wealth) as a function of the population under consideration. The proportion of the area under the diagonal (representing perfect equality) that is captured above the Lorenz curve represents how far the population pulls from perfect equality and is called the Gini Coefficient

---

[3]The area under the ROC curve (AUC) for 10-fold cross-validation experiments were 0.89 for gender, 0.72 for age, 0.72 for income, and 0.76 for education.

to be compared against the cumulative (in increasing shares) percentage of a corresponding population. The inequality present is illustrated by the degree to which the curve deviates from the straight diagonal ($y = x$) representative of perfect equality. There are two metrics that summarize Lorenz curves as a single statistic: (1) the Gini coefficient is defined as the proportion of the area under the line of perfect equality that is captured above the Lorenz curve, and (2) the Palma ratio, defined as the ratio of the share of the top 10% to the bottom 40% of users in the population.

Again, using wealth inequality as an example, if each individual in the population had a equal amount of wealth (perfect equality), the Lorenz curve would fall along the diagonal in Fig. 2, the Gini coefficient would be 0, and the Palma ratio would be 0.25 (10/40). Paired together, the Gini coefficient and Palma ratio provide a balanced understanding of the degree to which a Lorenz curve deviates from perfect equality. Gini coefficients are most sensitive to changes within the mid-range of the lorenz curve while the Palma is more sensitive to changes at the extremes.

1.5 Predicting User Reactions to Deceptive News

Discourse acts, or speech acts, can be used to identify the *use* of language within a conversation, e.g. agreement, question, or answer. In these studies, user reactions are classified as one of eight types of discourse acts analyzed in the context of social media discussions in previous work by Zhang et al.[49]: agreement, answer, appreciation, disagreement, elaboration, humor, negative reaction, or question, or as none of the given labels, denoted "other", using linguistically-infused neural network models [10].

1.6 Data

News Propagation and Influence from Deceptive Sources [11] Two datasets are used in this study. First, 11 million direct interactions (i.e. retweet and @mention) by almost 2 million Twitter users' who engaged with a set of 282 credibility-annotated news sources (using the approach described above) from January 2016 through January 2017. Second, a subset of these interactions for the 66,171 users who met all three of the following requirements: actively engaged (at least five times) with deceptive news sources, were identified as individual user accounts, and met the activity threshold of predictive models used to infer gender, age, income, and education demographics of the users. This dataset uses the fine-grained classifications of news sources identified as spreading trustworthy news, clickbait, conspiracy, propaganda, or disinformation.

Identifying and Understanding User Reactions to Deceptive and Trusted Social News Sources [10] User reactions to news sources were inferred for two popular platforms, Reddit and Twitter. The Reddit dataset comprises all Reddit posts submitted during the 13 month period from January 2016 through January 2017 that linked to domains associated with a previously annotated set of trustworthy versus deceptive news sources [11, 43] and the immediate comments (i.e. that directly responded to one of the posts). The Twitter dataset contains all tweets posted in the same 13 month period that directly @mentioned or retweeted content from one of these source's Twitter accounts. Coarse-grained news source classifications sets are used in this study: trustworthy, deceptive, and misinformation and disinformation.

How Humans versus Bots React to Deceptive and Trusted News Sources: A Case Study of Active Users [9] The dataset used in this study comprises a 431,771 tweets sample identified as English-content in the Twitter metadata of tweets posted between January 2016 and January 2017 that @mentioned or retweeted content from one of the annotated news sources (described above) also used for cross-platform and demographics-based engagement studies [10, 11]. This study focused on users who frequently interacted (at least five times) with deceptive news sources and considered fine-grained classifications of news sources. Each tweet was assigned a reaction type and user account type (bot or human) using the annotation processes described above – inferred via linguistically infused models [10] or based on botometer scores of users who authored each post [2].

2 *Who* Engages with (mis) and (dis)information?

Some studies model misinformation or rumor diffusion as belief exchange caused by influence from a users network, ego-network, or friends, e.g. the Tipping Model [35] and several previous studies have investigated the characteristics of users that spread or promote information as a way to identify those who spread rumors or disinformation [33]. For example, a 2015 study by Wu et al. [47] highlighted the type of user who shared content as one of their most important features in predictive models that were able to detect false rumors on Weibo with 90% confidence as quickly as 24 hours after the content was initially broadcast on the social network. Ferrara [3] found that users with high followings generated highly-infectious cascades for propaganda information. Recent work has also found that accounts spreading disinformation are significantly more likely to be automated accounts [36].

In this section, we focus on *who* engages with misinformation and disinformation and highlight key findings from several recent studies [9, 11, 21] related to user engagement with news sources categorized using the fine-grained classifications of: Trustworthy, Clickbait, Conspiracy, Propaganda, and Disinformation.

2.1   The Population Who Engage with Misinformation and Disinformation

Studies have identified that when an individual believes in one conspiracy theory, that individual is also likely to believe in others [12, 26]. At an aggregate level, one can consider whether this pattern might also hold for propagation or engagement with disinformation online – if a user engages or spreads mis and disinformation once, are they likely to engage again? if a user engages with news sources who publish one kind of deceptive content (e.g. clickbait), are they also likely to engage with another (e.g. intentional disinformation)? When investigated as a population as a whole, Glenski et al. [11] found that there were overlaps between populations of users engaging with news sources of varied degree of deception (illustrated in Fig. 3) but that the increased likelihood of sharing another type of deceptive news given that you engaged with another was not always reciprocal. For example, users who engage with news sources who spread clickbait and conspiracy theories are likely to also engage propaganda sources, but not the other way around.

Figure 4 highlights the degree to which engagement with news sources is evenly spread (or not) across the population who engage with news sources spreading trustworthy information versus mis- or disinformation. Unsurprisingly, disinformation sources are most highly retweeted from a small group of users that actively engage with those sources regularly. Effectively, a disproportionate amount of the engagement, promotion, or propagation of content published by news sources who were identified as spreading intentional disinformation from a subset of highly active, vocal users. Propaganda is the next most unevenly engaged with news, followed by trustworthy news, conspiracy, and clickbait.

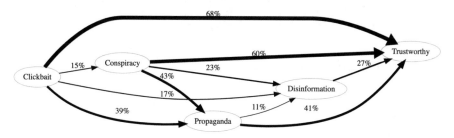

Fig. 3 Overlaps of users who engage with news sources across the spectrum of credibility as a directed graph for overlaps of at least 10% of users. Edges illustrate the tendency of users who engage with a news source spreading one type of deceptive content to also engage with a news source spreading another type of deceptive content. For example, the edge from Clickbait to Trustworthy illustrates that 68% of users who engage with news sources that spread clickbait, also engage with trustworthy news. Note: in total, 1.4 M users engaged with Trustworthy news sources, 19 k with Clickbait, 35.8 k with Conspiracy, 233.8 k with Propaganda, and 292.4 k with Disinformation

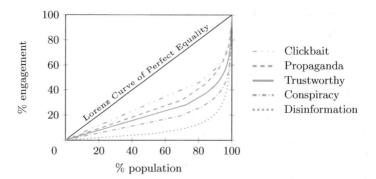

Fig. 4 Lorenz curves for inequality in engagement with news sources identified as spreading trustworthy news, clickbait, conspiracy theories, propaganda, and intentional disinformation. Legend (at right) is ordered from closest to furthest from the diagonal (representing engagement that is equally distributed across the population)

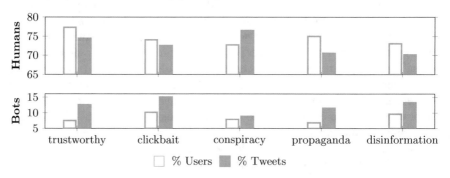

Fig. 5 Prevalence of humans (above) and bots (below) within tweets responding to news sources of varied credibility (% tweets, as solid bars) and within the populations of user accounts who authored the response-tweets (% users, as white-filled bars)

2.2 Automated Versus Manual Accounts

Glenski et al. [9] found that automated bot user accounts were responsible for approximately 9–15% of the direct responses to news sources spreading (mis) and (dis)information across all five fine-grained classifications of trustworthy news sources and news sources identified as spreading misinformation – clickbait, conspiracy, or propaganda – and intentional disinformation but only comprise around 7–10% of the users responsible for those response-tweets. Figure 5 illustrates the prevalence of automated (i.e. bots) versus manually-run (i.e. human) accounts among users who react to news sources spreading (dis)information within each category and the responses themselves (i.e. the percentage of tweets authored by user accounts inferred as automated versus manually run accounts).

Although news sources who spread conspiracy have the lowest presence of human users (72.8% of user accounts who authored reaction tweets), they have a disproportionately high proportion of reactions tweets authored by human-users (76.5% of tweets)—the highest proportion of human-authored reaction tweets across all five classes of news sources including the Trustworthy news sources which have the highest relative presence of human users. Interestingly, clickbait sources have the highest presence of bots with 10.17% of users identified as bots (who were responsible for 15.06% of the reaction-tweets) while news sources who spread disinformation have the second highest proportions of bots for users who reacted as well as reaction tweets posted.

2.3   Sockpuppets: Multiple Accounts for Deception

While bots are effective in spreading deceptive information at a fast speed and a large scale, the technology is not advanced enough to make their conversations and behavior believable as humans. This makes them barely effective in one-on-one conversations. Thus, bad actors adopt a smart strategy to deceive the audience: they create multiple accounts and operate them simultaneously to converse with the audience [21]. Kumar et al. showed that puppetmasters typically operate two or more 'sockpuppet' accounts, with the primary goal of deceiving others. These sockpuppet accounts typically support one another and create an illusion of magnified consensus. However, sometimes their strategies are more complex—instead of overtly supporting one another, some sockpuppet accounts oppose one another to create an illusion of argument. This attracts more attention and gets the audience involved as well. These crafty arguments are eventually used to influence people's opinions and deceive them.

Thus, the complex deceptive ecosystem created by the sockpuppets leads to increased attention to and the spread of false information and propaganda.

2.4   Demographic Sub-populations

When considering only the user accounts that frequently interacted with deceptive news sources on Twitter and the users' inferred demographics [11], the population was found to be primarily predicted to be male (96%), older (95%), with higher incomes (81%), college-educated (82%), and classified as "regular users" who followed more accounts than they had followers (59%), illustrated in Fig. 6. Although intuitively, this sample would not be expected to be a representative sample of Twitter users overall, the sample's majority demographic aligned with that found in a Pew Resarch Center survey conducted during the time period covered by the study – the Pew Research center survey found that 17% of Twitter users had a high school education or less, 38% were between 18 and 29 years old, and 47% were male [13] – although the study's sample was more heavily skewed towards the majority demographic than the Pew Research Center's findings.

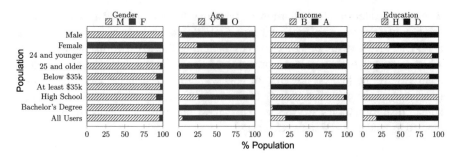

Fig. 6 Inferred demographics of users who frequently engage with deceptive news sources on Twitter [11]

Table 2 Inequality in user authorship of feedback to news sources who spread information (trustworthy news), misinformation (clickbait, conspiracy, and propaganda), and disinformation. Illustrated using the Palma Ratio – the ratio of feedback posted by the top 10% most active users to the 40% least active

		Trustworthy	Clickbait	Conspiracy	Propaganda	Disinformation
Gender	Male	4.47	1.72	6.58	3.58	25.06
	Female	3.08	1.40	1.44	3.04	18.03
Age	≤ 24	2.18	1.37	2.28	2.91	41.34
	≥ 25	4.48	1.72	6.58	3.58	18.15
Income	< $35k$	4.67	2.38	8.75	4.41	35.49
	≥ $35k$	4.35	1.66	6.00	3.39	8.99
Education	High school	4.09	2.19	8.88	4.17	35.52
	College	4.49	1.68	6.02	3.45	10.66
User Role	Follow	4.80	1.79	7.73	3.74	28.26
	Lead	4.00	1.65	5.37	3.42	21.29

There are significant differences in how equally users contribute to the feedback provided to news sources who spread information, misinformation, and disinformation online. We highlight the inequality in authorship of feedback (the extent to which a subset of highly active users contribute a disproportionate amount of the feedback within sub-populations by demographic) from Glenski et al. [11] in Table 2. Of note, the largest disparity in participation of feedback is in the sub-population of users inferred to be 24 years old or younger – the most active 10% of these users which provide feedback to disinformation sources via retweets or mentions author 41.34 as much as the least active 40% of users within this subpopulation. In contrast, the older sub-population (≥25 years old) have a much smaller palma ratio of 18.15. Overall, there is much greater inequality in participation of users who respond to news sources identified as spreading disinformation. Interestingly, the set of news sources which elicit the closest to uniform participation from responding users is clickbait, the least deceptive of the news sources who spread misinformation, rather than news sources identified as spreading trustworthy *information*.

3   *What* Kind of Feedback Do Users Provide?

In this section focusing on *what* kind of feedback users provide to news sources who spread information, misinformation, and disinformation, we highlight key findings from two recent studies [9, 10] related to the kinds of reactions (asking questions; expressing agreement, disagreement, or appreciation; providing answers; etc.) users post in response to social media news sources categorized using both the coarse-grained classifications of: Trustworthy, Deceptive, or Deceptive+Disinformation news sources [10] across two popular, and very different, social media platforms (Twitter and Reddit) and fine-grained classifications of: Trustworthy, Clickbait, Conspiracy, Propaganda, and intentional Disinformation [9] across user account characteristics (whether account is automated—i.e. a bot—or manually run).

3.1   *Across Multiple Platforms*

Glenski et al. [10] found that the predominant kinds of feedback elicited by any type of news source—from trustworthy sources sharing factual information without an intent to deceive the audience to deceptive news sources who spread intentional disinformation—across both Twitter and Reddit were answers, expressions of appreciation, elaboration on content posted by the news source, and questions. Figure 7 illustrates the distribution of these types of feedback, denoted reaction types, among Reddit comments (top plot) or tweets (bottom plot) responding to each category of news source (using the coarse classification as trustworthy versus deceptive or deceptive + disinformation) as a percentage of all comments/tweets

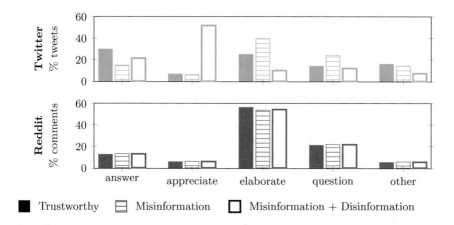

Fig. 7 Distributions of the five most frequently occurring ways in which users engage with news sources on Twitter (above) and Reddit (below) for coarse-grained partitions of trustworthy, misinformation, and misinformation + disinformation spreading news sources

reacting to sources of the given type (i.e. trusted, all deceptive, and deceptive excluding disinformation sources).

There were clear differences in the kinds of feedback posed to news sources on Twitter. As shown by the misinformation + disinformation bars, the misinformation news sources, when including disinformation-spreading sources, have a much higher rate of appreciation reactions and a lower rate of elaboration responses, compared to trustworthy news sources. Feedback from users towards disinformation spreading news sources are more likely to offer expressions of appreciation than elaboration. Differences are still significant ($p < 0.01$) but the trends reverse when the set of misinformation news sources do not include those that spread disinformation (including only those that spread clickbait, conspiracy, and propaganda). There is also an increase in the rate of question-reactions compared to trustworthy news sources when disinformation-spreading news sources are excluded from the set of deceptive news sources.

Feedback provided via user engagement on Reddit appears to follow a very similar distribution across different types of feedback for trustworthy versus misinformation/disinformation sources. However, Mann-Whitney U tests on Reddit-based user engagement still found that the illustrated differences between trusted and misinformation + disinformation news sources were statistically significant ($p < 0.01$)—regardless of whether we include or exclude disinformation sources. Posts that link to misinformation + disinformation sources have higher rates of expressions of appreciation and posing or answering questions while posts that link to trustworthy sources have higher relative rates of providing additional information or details via elaborations, expressions of agreement, and expressions of disagreement.

3.2   Across User-Account Characteristics

When the distributions of each class are compared, we find several key differences in what kind of feedback (i.e. reaction indicated from the primary discourse act of user response) is elicited. Conspiracy news sources have the highest relative rate of elaboration responses, i.e. *"On the next day, radiation level has gone up. [url]"* – with a more pronounced difference within the bot population – and the lowest relative rate of feedback in the manner of providing answers within the bot population but not within manually run accounts (i.e. human users). Clickbait news sources, on the other hand, have the highest relative rate of feedback where users provide answers and the lowest rate of where users pose questions across both populations of user account types (Fig. 8).

Conspiracy and propaganda news sources have higher rates within the population of manual accounts of accounts raising questions in response to the news sources than providing answers; manually run "human" accounts who respond to these types

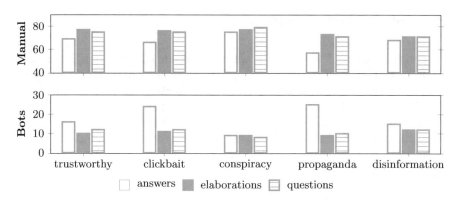

Fig. 8 Percentages of feedback of a given type (i.e. answers, elaborations, or questions) that were posted by manually run individual (above) and automated bot (below) user-accounts for each of the fine-grained classifications of news sources: those who spread trustworthy news, clickbait, conspiracy, propaganda, or disinformation

of news sources question the content posted by the source more often than they provide answers in response to a news source's posting. When reactions authored by bot-accounts are examined, there is a similar trend for conspiracy sources but a higher relative rate of answer reactions than question reactions to propaganda sources.

4 *How Quickly* Do Users Engage with (mis) and (dis)information?

Information diffusion studies have often used epidemiological models, originally formulated to model the spread of disease within a population, in the context of social media [16, 41, 48]. For example, Tambuscio et al. [41] used such a model to determine a threshold of fact-checkers needed to eliminate a hoax. In this context, users are *infected* when they spread information to other users. A recent study by Vosoughi et al. [44] found that news that was fact-checked (post-hoc) and found to be false had spread faster and to more people than news items that were fact-checked and found to be true. In this section, we highlight key findings on the speed at which users react to content posted by news sources of varying credibility and comparative analyses of the delays of different types of responses. By contrasting the speed of reactions of different types, from different types of users (bot and human), and in response to sources of varying credibility, one is able to determine whether deceptive or trusted *sources* have slower immediate share-times overall or within combinations of classes of user account or news sources.

4.1 Across Multiple Platforms

In [10], Glenski et al. examine the speed and volume of user engagement with social news using coarse-grained partitioning of sources as trustworthy or deceptive (e.g. news sources that spread a variety of disinformation). A key finding was the differences in the pace and longitude of engagement with the same deceptive news sources across differing social platforms: Twitter and Reddit. The duration of engagement with content across trustworthy and deceptive news sources alike was found to be typically more prolonged for engagement with information spread on Twitter compared to Reddit. Intuitively, this could be due to the different manner in which users engage with content in general when using one platform versus another. Users are able to pinpoint specific *users* (or news source accounts) to follow, regularly consume content from, or easily engage with on Twitter whereas users "follow" topics, areas of interest, or communities of users through the Reddit mechanism of subscribing to subreddits. While news sources have content spreading across both, there is a greater difficulty to consistently engage with a single news sources content over time on Reddit.

Cumulative density function plots for three means of engagement are illustrated in Fig. 9 for the sets of trustworthy, misinformation, and misinformation +

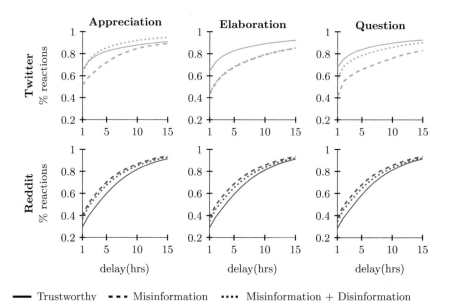

Fig. 9 Cumulative density function plots for three means of engagement (where users express *appreciation* towards the news source, *elaborate* on content published by the news source, or *question* the content published by a news source) for the sets of trustworthy, misinformation, and misinformation + disinformation news sources when users engaged via the Twitter (above) and Reddit (below) platforms

disinformation spreading news sources when users engaged via the Twitter and Reddit platforms. In addition to the differences in scale of duration of engagement, *engagement with trustworthy social news sources are less heavily concentrated within the first 12 to 15 h after content is initially published by the social news source on Reddit whereas the opposite is found on Twitter. While Twitter social news sources may have a larger range of delays before a user engages with content, they are also more heavily skewed with larger concentrations immediately following a source's publication of content* ($p < 0.01$).

If delays in providing feedback are examined using more fine-grained classifications of misinformation [11], the populations of users who provide feedback to news sources on Twitter to news sources who spread trustworthy information, conspiracy, and disinformation news have similarly short delays between the time when a news source posts new content on Twitter and when a user provides feedback via @mentioning or retweeting the news source. However, delays are significantly longer for news sources identified as spreading clickbait and propaganda misinformation ($p < 0.01$).

4.2 Across User-Account Characteristics

Next, we highlight the speed with which bot and human users react to news sources [9]. As would be expected, this study found that response activity is heavily concentrated in the window of time soon after a news source posts when considering any combination of type of information being spread or feedback being provided. Mann Whitney U tests that compared distributions of response delays found that manually-run accounts will pose questions and provide elaborations of information posted by news sources those that spread clickbait faster than automated bot accounts do ($p < 0.01$); There is a heavier concentration (at least 80%) of reactions from manually-run accounts that have response delays with at most a 6 h delay compared to automated bot accounts that have approximately 60–70% of their elaboration and question based responses falling within that initial 6 h window, shown in Fig. 10.

There are similar trends for all the other combinations of feedback provided to and type of information spread by news sources with a few notable exceptions: (1) automated bot accounts provide answer-responses to news sources identified as spreading propaganda content with significantly shorter delays than manually-run accounts ($p < 0.01$) and (2) MWU tests comparing sub-populations of automated and manual accounts authoring feedback providing answers to news sources who spread either clickbait or disinformation were not found to differ with statistical significance.

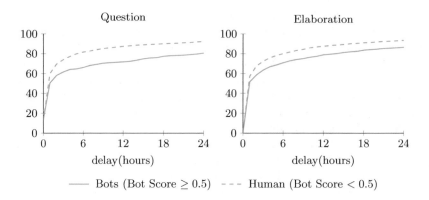

Fig. 10 Cumulative distribution function (CDF) plots of the volume by the delay between when a news source identified as spreading clickbait posted content and when a response that posed a question (left) or elaboration (right) was posted for bots and human user accounts, using a step size of one day

Table 3 Users by demographics who respond faster to news sources identified as spreading *trustworthy* information, several classes of misinformation (*clickbait, conspiracy, propaganda*), and *disinformation*. A dash (—) indicates no significant differences were found between sub-populations for a given demographic with all other results being statistically significant (MWU $p < 0.01$)

	Trustworthy	Clickbait	Conspiracy	Propaganda	Disinformation
Gender	Male	—	Male	Male	Male
Age	≥ 25	—	≤ 24	≤ 24	≤ 24
Income	$< 35k$	$< 35k$	$< 35k$	$< 35k$	$< 35k$
Education	High school	High school	High school	High school	High school
Role	Leader	Follower	Follower	Leader	Follower

4.3 Demographic Sub-populations

In Table 3, we highlight the speed of response comparisons by demographic sub-population from [11]. Older users were found to retweet news sources identified as spreading trustworthy news more quickly than their younger counterparts but slower to share all other types – that is, younger users (≤ 24 years old) engage with the deceptive news sources (misinformation and disinformation spreading alike) more quickly, sooner after the news source is posting content. Users inferred to have only high school education engage faster than those with a college education across the board. Except for comparisons between predicted gender or age brackets for clickbait sources, there are statistically significant differences in delays for all information type and demographic combinations.

5 Discussion and Conclusions

In this chapter, we have highlighted key findings of several recent studies that examined the human element of the digital deception ecosystem and news cycles— the audience who engage with, spread, and consume the misinformation and disinformation present on the online social platforms that society has come to rely on for quick and convenient consumption of information, opinion, and news. Framed to answer each of our three key research questions, we have presented the key findings of several recent studies and how the results pair together to present a comprehensive understanding of user engagement with multiple scales or resolutions of deception (from coarse to fine-grained credibility annotations).

However, in each of the studies referenced above, we have analyzed audience engagement from the position of knowing whether news sources or content is deceptive or trustworthy. Often, if not always, individual users are not given such clear labels of deception versus not. Rather, they are faced with the opposite, where deceptive and trustworthy content and news sources alike portray themselves as trustworthy. A key premise of studying the audience reaction to misinformation and disinformation is that they should be able identify false information when they come across it in social media. But how effective are readers in identifying false information? To answer this question, Kumar et al. [23] conducted a human experiment using hoax articles on Wikipedia as disinformation pieces and non-hoax articles as non-deceptive pieces.

Hoax articles on Wikipedia contain completely fabricated information and are created with the intention of deceiving others. Identifying them on Wikipedia requires a meticulously manual process that guarantees the ground truth. In a human experiment, Kumar et al. showed a pair of articles to Mechanical Turk workers— one article was a hoax article and another was a non-hoax article—and the workers were told to identify the hoax article. In this scenario, random guess would yield a 50% accuracy while the workers got the answer correct 66% of the times. This shows that humans are able to identify false information better than random though they are not perfect. Analysis of their mistakes showed that well-formatted, long, and well-referenced hoax articles fooled humans into thinking it is true. This shows that humans *can* be able to identify false information when they come across it, as shown in this setting of Wikipedia content. However, the real power comes when leveraging feedback at a large scale from a sizeable audience in social media.

Similarly, Karduni et al. [20] conducted human experiments to study user decision-making processes around misinformation on Twitter and how uncertainty and confirmation bias (the tendency to ignore contradicting information) affect users decision-making. The authors developed visual analytic system – Verifi[4] designed provide users with the ability to characterize and distinguish misinformation from legitimate news. Verifi explicitly presents a user with the cues to make decisions

---

[4]https://verifi.herokuapp.com/

about the veracity of news media sources on Twitter including account-level temporal trends, social network and linguistic features e.g., biased language, subjectivity, emotions etc. The authors then used Verifi to measure how users assess the veracity of the news media accounts on Twitter (focusing on textual content rather than images) and what role confirmation bias plays in this process. Their analysis shows that certain cues significantly affected users decisions about the veracity of news sources more than others, for example specific named entities, fear and negative language and opinionated language. However, similar to Kumar et al. study, user accuracy rate ranges between 54% and 74% depending on different experimental conditions.

Verifi2 [19], a visual analytic system that enables users to explore news in an informed way by presenting a variety of factors that contribute to its veracity. It allows to contrast (1) language used by real and suspicious news sources, (2) understand the relationship between different news sources, (3) understand top names entities, and (4) compare real vs. suspicious news sources preferences on images. The authors conduct interviews with experts in digital media, communications, education, and psychology who study misinformation in order to help real users make decisions about misinformation in real-world scenarios. All of their interviewees acknowledged the challenge in defining misinformation, as well as the complexity of the issue which involves both news outlets with different intents, as well as audiences with different biases. Finally, Verifi2 expert users suggested to define a spectrum of trustworthiness rather than binary classes (real vs. suspicious news sources), and identified the potentials for Verifi2 to be used in scenarios where experts educate individuals about differences between real and suspicious sources of news.

A well-rounded understanding of existing patterns, trends, and tendencies of user engagement is a necessary basis for the development of effective strategies to defend against the evolving threat of digital deception. Key findings highlighted here in the context of multiple studies at varied resolutions of credibility of information or sources, user account characteristics, and social platforms under consideration can be used to inform models and simulations of (dis)information spread within and across communities of users, social platforms, geolocations, languages, and types of content. Further, they can be used to advise direct interventions with individuals or groups of users to improve their manual detection skills. Some open challenges include how to effectively combine feedback from large audience in real-time and how to improve detection of complex multimedia disinformation using audience feedback.

Acknowledgements This work was supported in part by the Laboratory Directed Research and Development Program at Pacific Northwest National Laboratory, a multiprogram national laboratory operated by Battelle for the U.S. Department of Energy. This research was also supported by the Defense Advanced Research Projects Agency (DARPA), contract W911NF-17-C-0094. The U.S. Government is authorized to reproduce and distribute reprints for Governmental purposes notwithstanding any copyright annotation thereon. The views and conclusions contained herein are those of the authors and should not be interpreted as necessarily representing the official policies or endorsements, either expressed or implied, of DARPA or the U.S. Government. This work has also been supported in part by Adobe Faculty Research Award , Microsoft, IDEaS, and Georgia Institute of Technology.

References

1. Bikhchandani, S., Hirshleifer, D., Welch, I.: A theory of fads, fashion, custom, and cultural change as informational cascades. J. Polit. Econ. **100**(5), 992–1026 (1992). https://doi.org/10.2307/2138632, http://www.jstor.org/stable/2138632
2. Davis, C.A., Varol, O., Ferrara, E., Flammini, A., Menczer, F.: Botornot: a system to evaluate social bots. In: Proceedings of the 25th International Conference Companion on World Wide Web, pp. 273–274. International World Wide Web Conferences Steering Committee (2016)
3. Ferrara, E.: Contagion dynamics of extremist propaganda in social networks. Inf. Sci. **418**, 1–12 (2017)
4. Ferrara, E.: Disinformation and social bot operations in the run up to the 2017 french presidential election. First Monday **22**(8) (2017). https://doi.org/10.5210/fm.v22i8.8005, http://journals.uic.edu/ojs/index.php/fm/article/view/8005
5. Gabielkov, M., Ramachandran, A., Chaintreau, A., Legout, A.: Social clicks: what and who gets read on twitter? ACM SIGMETRICS Perform. Eval. Rev. **44**(1), 179–192 (2016)
6. Glenski, M., Pennycuff, C., Weninger, T.: Consumers and curators: browsing and voting patterns on reddit. IEEE Trans. Comput. Soc. Syst. **4**(4), 196–206 (2017)
7. Glenski, M., Weninger, T.: Predicting user-interactions on reddit. In: Proceedings of the IEEE/ACM International Conference on Advances in Social Networks Analysis and Mining (ASONAM). IEEE/ACM (2017)
8. Glenski, M., Weninger, T.: Rating effects on social news posts and comments. ACM Trans. Intell. Syst. Technol. (TIST) **8**(6), 1–9 (2017)
9. Glenski, M., Weninger, T., Volkova, S.: How humans versus bots react to deceptive and trusted news sources: a case study of active users. In: Proceedings of the IEEE/ACM International Conference on Advances in Social Networks Analysis and Mining (ASONAM). IEEE/ACM (2018)
10. Glenski, M., Weninger, T., Volkova, S.: Identifying and understanding user reactions to deceptive and trusted social news sources. In: Proceedings of the 56th Annual Meeting of the Association for Computational Linguistics (Volume 2: Short Papers), vol. 2, pp. 176–181 (2018)
11. Glenski, M., Weninger, T., Volkova, S.: Propagation from deceptive news sources who shares, how much, how evenly, and how quickly? IEEE Trans. Comput. Soc. Syst. **5**(4), 1071–1082 (2018)
12. Goertzel, T.: Belief in conspiracy theories. Polit. Psychol. **15**(4), 731–742 (1994)
13. Gottfried, J., Shearer, E.: News use across social media platforms 2016. Pew Research Center (2016). http://www.journalism.org/2016/05/26/news-use-across-social-media-platforms-2016/
14. Hasson, U., Simmons, J.P., Todorov, A.: Believe it or not: on the possibility of suspending belief. Psychol. Sci. **16**(7), 566–571 (2005)
15. Hirshleifer, D.A.: The blind leading the blind: social influence, fads and informational cascades. In: Ierulli, K., Tommasi, M. (eds.) The New Economics of Human Behaviour, chap 12, pp. 188–215. Cambridge University Press, Cambridge (1995)
16. Jin, F., Dougherty, E., Saraf, P., Cao, Y., Ramakrishnan, N.: Epidemiological modeling of news and rumors on twitter. In: Proceedings of the Seventh Workshop on Social Network Mining and Analysis, p. 8. ACM (2013)
17. Kakwani, N.C., Podder, N.: On the estimation of lorenz curves from grouped observations. Int. Econ. Rev. **14**(2), 278–292 (1973)

18. Karadzhov, G., Gencheva, P., Nakov, P., Koychev, I.: We built a fake news & click-bait filter: what happened next will blow your mind! In: Proceedings of the International Conference on Recent Advances in Natural Language Processing (2017)
19. Karduni, A., Cho, I., Wesslen, R., Santhanam, S., Volkova, S., Arendt, D.L., Shaikh, S., Dou, W.: Vulnerable to misinformation? Verifi! In: Proceedings of the 24th International Conference on Intelligent User Interfaces, pp. 312–323. ACM (2019)
20. Karduni, A., Wesslen, R., Santhanam, S., Cho, I., Volkova, S., Arendt, D., Shaikh, S., Dou, W.: Can you verifi this? studying uncertainty and decision-making about misinformation using visual analytics. In: Twelfth International AAAI Conference on Web and Social Media (2018)
21. Kumar, S., Cheng, J., Leskovec, J., Subrahmanian, V.: An army of me: sockpuppets in online discussion communities. In: Proceedings of the 26th International Conference on World Wide Web, pp. 857–866. International World Wide Web Conferences Steering Committee (2017)
22. Kumar, S., Shah, N.: False information on web and social media: a survey. arXiv preprint arXiv:1804.08559 (2018)
23. Kumar, S., West, R., Leskovec, J.: Disinformation on the web: impact, characteristics, and detection of wikipedia hoaxes. In: Proceedings of the 25th International Conference on World Wide Web, pp. 591–602. International World Wide Web Conferences Steering Committee (2016)
24. Kwon, S., Cha, M., Jung, K.: Rumor detection over varying time windows. PLoS One $12(1)$, e0168344 (2017)
25. Kwon, S., Cha, M., Jung, K., Chen, W., Wang, Y.: Prominent features of rumor propagation in online social media. In: Proceedings of the 13th International Conference on Data Mining (ICDM), pp. 1103–1108. IEEE (2013)
26. Lewandowsky, S., Oberauer, K., Gignac, G.E.: Nasa faked the moon landing – therefore,(climate) science is a hoax: an anatomy of the motivated rejection of science. Psychol. Sci. $24(5)$, 622–633 (2013)
27. Lorenz, J., Rauhut, H., Schweitzer, F., Helbing, D.: How social influence can undermine the wisdom of crowd effect. Proc. Natl. Acad. Sci. $108(22)$, 9020–9025 (2011)
28. Matsa, K.E., Shearer, E.: News use across social media platforms 2018. Pew Research Center (2018). http://www.journalism.org/2018/09/10/news-use-across-social-media-platforms-2018/
29. Mitra, T., Wright, G.P., Gilbert, E.: A parsimonious language model of social media credibility across disparate events. In: Proceedings of the ACM Conference on Computer Supported Cooperative Work and Social Computing (CSCW), pp. 126–145. ACM (2017)
30. Muchnik, L., Aral, S., Taylor, S.J.: Social influence bias: a randomized experiment. Science $341(6146)$, 647–651 (2013)
31. Qazvinian, V., Rosengren, E., Radev, D.R., Mei, Q.: Rumor has it: identifying misinformation in microblogs. In: Proceedings of the Conference on Empirical Methods in Natural Language Processing (EMNLP), pp. 1589–1599. Association for Computational Linguistics (2011)
32. Rashkin, H., Choi, E., Jang, J.Y., Volkova, S., Choi, Y.: Truth of varying shades: analyzing language in fake news and political fact-checking. In: Proceedings of the 2017 Conference on Empirical Methods in Natural Language Processing, pp. 2921–2927 (2017). https://aclanthology.info/papers/D17-1317/d17-1317
33. Rath, B., Gao, W., Ma, J., Srivastava, J.: From retweet to believability: utilizing trust to identify rumor spreaders on twitter. In: Proceedings of the IEEE/ACM International Conference on Advances in Social Networks Analysis and Mining (ASONAM) (2017)
34. Rubin, V.L., Conroy, N.J., Chen, Y., Cornwell, S.: Fake news or truth? Using satirical cues to detect potentially misleading news. In: Proceedings of NAACL-HLT, pp. 7–17 (2016)
35. Schelling, T.C.: Micromotives and macrobehavior. WW Norton & Company, New York (2006)
36. Shao, C., Ciampaglia, G.L., Varol, O., Flammini, A., Menczer, F.: The spread of fake news by social bots. arXiv preprint arXiv:1707.07592 (2017)
37. Starbird, K.: Examining the alternative media ecosystem through the production of alternative narratives of mass shooting events on twitter. In: Proceedings of the 11th International AAAI Conference on Web and Social Media (ICWSM). AAAI (2017)

38. Starbird, K., Maddock, J., Orand, M., Achterman, P., Mason, R.M.: Rumors, false flags, and digital vigilantes: misinformation on twitter after the 2013 Boston marathon bombing. iConference 2014 Proceedings (2014)
39. Street, C.N., Masip, J.: The source of the truth bias: Heuristic processing? Scand. J. Psychol. **56**(3), 254–263 (2015)
40. Takahashi, B., Tandoc, E.C., Carmichael, C.: Communicating on twitter during a disaster: an analysis of tweets during typhoon haiyan in the philippines. Comput. Human Behav. **50**, 392–398 (2015)
41. Tambuscio, M., Ruffo, G., Flammini, A., Menczer, F.: Fact-checking effect on viral hoaxes: a model of misinformation spread in social networks. In: Proceedings of the 24th International Conference on World Wide Web, pp. 977–982. ACM (2015)
42. Volkova, S., Bachrach, Y.: Inferring perceived demographics from user emotional tone and user-environment emotional contrast. In: Proceedings of the 54th Annual Meeting of the Association for Computational Linguistics (Volume 1: Long Papers), vol. 1 (2016)
43. Volkova, S., Shaffer, K., Jang, J.Y., Hodas, N.: Separating facts from fiction: linguistic models to classify suspicious and trusted news posts on Twitter. In: Proceedings of the 55th Annual Meeting of the Association for Computational Linguistics (Volume 2: Short Papers), vol. 2, pp. 647–653 (2017)
44. Vosoughi, S., Roy, D., Aral, S.: The spread of true and false news online. Science **359**(6380), 1146–1151 (2018). https://doi.org/10.1126/science.aap9559
45. Wang, W.Y.: "Liar, liar pants on fire": a new benchmark dataset for fake news detection (2017)
46. Weninger, T., Johnston, T.J., Glenski, M.: Random voting effects in social-digital spaces: a case study of reddit post submissions. In: Proceedings of the 26th ACM Conference on Hypertext & Social Media, pp. 293–297. HT '15, ACM, New York (2015). https://doi.org/10.1145/2700171. 2791054
47. Wu, K., Yang, S., Zhu, K.Q.: False rumors detection on Sina Weibo by propagation structures. In: Proceedings of the 31st International Conference on Data Engineering (ICDE), pp. 651–662. IEEE (2015)
48. Wu, L., Morstatter, F., Hu, X., Liu, H.: Mining misinformation in social media. In: Big Data in Complex and Social Networks, CRC Press, pp. 123–152 (2016)
49. Zhang, A., Culbertson, B., Paritosh, P.: Characterizing online discussion using coarse discourse sequences. In: Proceedings of the 11th International AAAI Conference on Web and Social Media (ICWSM). AAAI (2017)

Characterization and Comparison of Russian and Chinese Disinformation Campaigns

David M. Beskow and Kathleen M. Carley

Abstract While substantial research has focused on social bot classification, less computational effort has focused on repeatable bot characterization. Binary classification into "bot" or "not bot" is just the first step in social cybersecurity workflows. Characterizing the malicious actors is the next step. To that end, this paper will characterize data associated with state sponsored manipulation by Russia and the People's Republic of China. The data studied here was associated with information manipulation by state actors, the accounts were suspended by Twitter and subsequently all associated data was released to the public. Of the multiple data sets that Twitter released, we will focus on the data associated with the Russian Internet Research Agency and the People's Republic of China. The goal of this paper is to compare and contrast these two important data sets while simultaneously developing repeatable workflows to characterize information operations for social cybersecurity.

Keywords Bot characterization · Social cybersecurity · Disinformation · Information operations · Strategic competition · Propaganada · Exploratory data analysis · Internet memes

1 Introduction

State and non-state actors leverage information operations to create strategic effects in an increasingly competitive world. While the art of influence and manipulation dates back to antiquity, technology today enable these influence operations at a scale and sophistication unmatched even a couple decades ago. Social media platforms have played a central role in the rise of technology enabled information warfare. As state and non-state actors increasingly leverage social media platforms as central to

D. M. Beskow (✉) · K. M. Carley
School of Computer Science, Carnegie Mellon University, Pittsburgh, PA, USA
e-mail: dbeskow@andrew.cmu.edu; kathleen.carley@cs.cmu.edu

© Springer Nature Switzerland AG 2020
K. Shu et al. (eds.), *Disinformation, Misinformation, and Fake News in Social Media*,
Lecture Notes in Social Networks, https://doi.org/10.1007/978-3-030-42699-6_4

Table 1 List of Datasets that Twitter has released in association of state sponsored information manipulation

Year-Month	Country	Tweets	Users
2018-10	Russia	9,041,308	3,667
2019-01	Russia	920,761	361
2019-01	Bangladesh	26,212	11
2019-01	Iran	4,671,959	2,496
2019-01	Venezuela	8,950,562	987
2019-04	Ecquador	700,240	787
2019-04	Saudi Arabia	340	6
2019-04	Spain	56,712	216
2019-04	UAE	1,540,428	3,898
2019-04	Venezuela	1,554,435	611
2019-06	Russia	2,288	3
2019-06	Iran	4,289,439	4,238
2019-06	Catalonia	10,423	77
2019-08	China	3,606,186	890
2019-09	China	10,241,545	4324

their ongoing information and propaganda operations, the social media platforms themselves have been forced to take action.

One of the actions that Twitter took is to suspend accounts associated with state sponsored propaganda campaigns and then release this data to the public for analysis and transparency. So far they have only released data associated with state sponsored manipulation and not other actor types. A summary of the data that they released is provided in Table 1 below. The largest and most prominent of these is the data associated with the Russian Internet Research Agency (IRA) and the Chinese data. The IRA data includes a well documented information campaign to influence an election and otherwise cause division in the United States, and the Chinese data is associated with information manipulation around the Hong Kong protests.

Our analysis of the Chinese and IRA data is a means for us to begin developing repeatable ways to characterize malicious online actors. Our experience is that social cybersecurity analysts often use a supervised machine learning algorithm to conduct their initial triage of a specific social media stream, say a stream related to an election event. This supervised model will often label tens of thousands of accounts as likely automated/malicious, which is still too many to sift through manually. While there are ways for an analyst to prioritize this list (for example finding the intersection of the set of likely bots with the set of influential actors measured with eigenvector centrality), it would be nice to characterize these malicious actors in a richer way than binary classification of "bot" or "not". This paper, using the IRA and Chinese data to illustrate, will pave the way for future research and tools that will provide a comprehensive bot-labeling workflow for characterizing malicious online actors.

The IRA data that we will study in this paper is the original data set that Twitter released under their then nascent elections transparency effort. This release was spurred by the fall-out after the 2016 US election and increasing evidence of Russian manipulation. The data has been studied as part of the Mueller Special Counsel investigation as well as several independent analysis conducted on behalf of the US Senate.

The Chinese data was produced from behind China's firewall and based on the IP addresses associated with the activity Twitter believes was produced by the People's Republic of China or a sanctioned proxy. This manipulation was attempting to change the narrative of the Hong Kong protest both for the residents of Hong Kong as well as the broader international community.

Before we spend some time going into a deeper comparison of these two data sets, we acknowledge that at a macro level they are very different because the target events are vastly different. In the case of the Russian IRA data, they were attempting to create a change in a foreign election on the other side of the world. In the Chinese case, they were largely trying to control the narrative of domestic events evolving inside their own borders. Acknowledging this macro level difference will shed some light on the other differences we uncover in this paper.

In addition to analyzing the *core* data that Twitter released to the public, we also collected additional data on all accounts that are mentioned, retweeted, replied to, or otherwise associated with the *core* data. This additional data was collected with the Twitter REST API, and throughout this paper we will refer to it as the *periphery* data. Note that this *periphery* data includes both malicious and non-malicious accounts. The malicious accounts have not been suspended by Twitter, and are either continuing to conduct information warfare or are in a dormant state waiting to be activated. The non-malicious accounts are accounts that became associated with the *core* data through a mention, retweet, or reply. These are often online actors that are either amplified or attacked in the information operation, or they could be innocent bystanders that bots and trolls mention in an attempt to build a following link so that they can influence them. Note that at the end of this paper we will attempt to estimate the number of accounts in the *periphery* data that are malicious and still active.

While several papers and reports as well as news articles have explored each of these data sets individually, as of the time of this writing we have not found a paper or report that expressly compares them. In conducting this research, our goal in order of priority is to:

1. Develop repeatable workflows to characterize information operations
2. Compare and contrast Russian and Chinese approaches to influence and manipulation of Twitter
3. Build on existing analysis of these unique data sets and the events and manipulation they are associated with

In order to characterize and then compare and contrast these data sets, we will develop and illustrate the use of social cybersecurity analytics and visualization. In this paper we will specifically focus on visual network analysis, new geographic analysis using flag emojis, temporal analysis of language and hashtag market share, bot analysis using several supervised machine learning models, meme analysis of image memes, and analysis of state sponsored media involvement. We will then finish up by analyzing and discussing the number of accounts in the periphery data that are still conducting or supporting state sponsored information manipulation. Research such as this is key for threat assessment in the field of social cybersecurity [19].

2   Literature Review

Several reports and research papers have explored the data that Twitter released relative to the Russian/IRA and Chinese information operations. These are discussed below.

2.1   Russia Internet Research Agency Data

Russia's Internet Research Agency (IRA) is a St. Petersburg based company that conducts information operations on social media on behalf of the Russian government and businesses. The company began operations in 2013 and has trained and employed over 1000 people [12].

The IRA data has had more time and research effort than the newer Chinese manipulation data. Even before Twitter released the data to the public they allowed several research organizations an early analysis to accompany the release. Notable among these preliminary and largely exploratory analysis is the research by the Digital Forensic Labs [17].

The Special Investigation "Mueller" report, released on April 18, 2019, detailed the IRA operations [16]. The 443 page report contains 16 pages dedicated to IRA manipulation of information surrounding the 2016 US Presidential election. The manipulation detailed in the redacted report includes organization of grassroots political efforts, use of accounts masquerading as grass roots political efforts. The report indicates that the IRA accounts posed as anti-immigration groups, Tea Party activists, Black lives matter activists, LGBTQ groups, religious groups (evangelical or Muslim groups), as well as other political activists. It also detailed the methods used and organization of personnel against these methods. Two IRA employees received visas and traveled to the United States in order to better understand the social, cultural, and political cultures. IRA employees operated accounts initially focused on Twitter, Facebook, and Youtube but eventually including Tumblr and Instagram accounts. It also details the purchase of advertisements. It details a separate bot network that amplified IRA inauthentic user content. It noted that celebrities, politicians, and news outlets quoted, retweeted, or otherwise spread IRA messaging. The report outlines throughout the 16 pages how messaging for Trump was positive and supportive while the messaging for Clinton was negative. The IRA was also central to the February 2018 indictment of 13 Russian nationals by Special Counsel Robert Mueller [1].

The second report regarding the IRA was conducted by New Knowledge at the request of the US Senate Select Committee on intelligence (SSCI) and focused on Facebook, Instagram, Twitter, Youtube, Google+, Gmail, and Google Voice involving the IRA. The report also shows some evidence of IRA activity on Vine, Gab, Meetup, VKontakte, and LiveJournal. The data that Twitter provided to New Knowledge was roughly the same data that was released to the public, but was not hashed and contained IP address and other information. This highlights the IRA switch from Facebook/Twitter to Instagram following their negative publicity. It highlighted that Instagram outperformed Facebook, highlighting the importance of images and memes in information operations. Like the Mueller report it highlights targeted communities. It also discusses voter suppression operations, such as encouraging voters to vote for a third candidate, stay home on election day, or false advertisements for voting on Twitter. In addition to highlighting pro-Trump and anti-Clinton campaigns, it also highlights activity meant to divide, such as secessionist messaging. It then conducts temporal analysis, URL analysis, and other content analysis. They highlight some of the tactics, branding, and recruitment. It also highlights the IRA's attacks against Republican primary candidates. They conduct extensive analysis of the memetic warfare. They highlight the IRA tactic of amplifying conspiracy theories. Finally, they thoroughly highlight efforts to divide America through secession ("if Brexit, why not Texit"). To summarize their analysis was primarily content, strategy, and effects across a sophisticated campaign that targeted Black, Left, and Right leaning groups [12].

The Computational Propaganda Project, like New Knowledge, was provided data by the US Senate Select Committee on Intelligence, to include the Twitter IRA data. In addition to temporal analysis, categorical analysis, target population identification, limited network analysis, hashtag and content analysis, It focused on cross platform activity [14].

Several other notable research efforts on the IRA include Arian Chen's lengthy New York Times Article entitled "The Agency" which details how the IRA organizes false alarms such as their Columbian Chemicals Explosion Hoax and the Ebola virus hoax [9]. Badawy et al conducts research of the 2016 IRA data and analyzes to what extent the effort supported the political left versus the political right [2], and is probably the closest article to the effort that we propose. Note that the Badawy effort only focuses on IRA data, and does not include any discussion of the Chinese data.

2.2   Chinese Manipulation of Hong Kong Narrative

In August 2019 Twitter released data associated with information and platform manipulation by the Chinese government around the Hong Kong protests. Twitter claims this was a state-backed information operation. As evidence for this claim, they point to the fact that all of the activity and the associated IP addresses on the suspended accounts originated from within the People's Republic of China (PRC) even though Twitter is blocked by the PRC (i.e. China's 'Great Firewall'). While some users in China access Twitter through VPNs, the nature of VPNs means the IP addresses aren't from within the PRC. Twitter suspended the accounts for violating terms of service [18]. Censorship, while well documented, is difficult to measure [13].

The China data has had limited reporting on it. This is partially because it is newer, and also because it is harder to put together a cohesive picture of the data. Any cursory exploratory data analysis will often leave the researcher puzzled. Multiple posts on social media and elsewhere express this puzzlement. This is because the highest languages in the data are Indonesian, Arabic, and English, not Chinese. The most common hashtag is PTL ("Praise the Lord"). A substantial part of the data appears to involve an escort service or prostitution ring in Las Vegas, Asia and possibly elsewhere. It is only after extensive analysis that we will walk through in this report that the true nature of the data becomes evident.

While there are limited reporting on this data, we do want to call attention to the most thorough analysis we've found to date. The most comprehensive analysis we've found was conducted by Uren et al at the Australian Strategic Policy Institute [22]. This research highlights that these accounts attacked political opponents of the Communist Party of China (CPC) even before they began influencing the events in Hong Kong. Some of the primary conclusions of the report is that the Chinese approach appears reactionary and somewhat haphazard. They did not embed in virtual groups and slowly build influence, but rather generated simple spam that supported their messaging. This report does go into extensive temporal and geographic analysis that we will at times enhance but not duplicate. They do highlight that the lack of sophistication may because it was outsourced to a contractor or because the government agency overseeing the operation lacked a full understanding of information operations. This report also highlights the fact that many of these accounts appear to be purchased at some point in their history. The authors show that 630 tweets contain phrases like 'test new owner', 'test', 'new own', etc. which are commonly used to show that a given account has come under new ownership.

3 Data

Twitter is a core platform for the global conversation, providing an open market for opinions and beliefs. By 2014 Twitter surpassed Facebook citations in the New York Times and by 2016 the New York Times cited Twitter more than twice as much as Facebook [23]. Online media often include Twitter posts of celebrities, politicians, and other elites in their content. To some extent, Twitter captures more of the global conversation (particularly in the West) while Facebook captures more of the local and topical conversations. Given this important opinion market, numerous

Table 2 Summary of data

	IRA		China	
	Core	Periphery	Core	Periphery
Tweets	9,041,308	47,741,450	3,606,186	32,616,654
Users	3,667	667,455	890	20,4145
Top 5 languages	ru,en,de,uk,bg	en,ru,es,de,ar	in,ar,en,pt,zh	en,ar,pt,in,es

actors attempt to market their ideas and at times manipulate the marketplace for their benefit.

As mentioned above, the data is divided into the *core* data that Twitter released, as well as the *periphery* data that was associated with the *core* data. The *periphery* data includes any account that was mentioned, replied to, or retweeted by the *core* data. For every account in the periphery data, we collected the associated timeline (up to last 200 tweets). A summary of the *core* and *periphery* data sets is provided in Table 2.

4   Characterization and Comparison

4.1   Network

Information operations by their very nature manipulate narratives and networks. In order to understand and characterize them, we must understand the network that they are embedded in. To do this, we used the core data that was suspended and released by Twitter, and developed a network of communications. Links in the network represent one of the directed communication actions a user can take on Twitter, namely mention, reply, and retweet. These are communication links, not friend/following links. Nodes in this graph include both *core* and *periphery* accounts.

The networks are seen in Fig. 1a, with nodes colored by their most recent language. In the case of the Russian IRA data, we see clear lines of effort in Russian and English. When we zoom in on some of the Russian language clusters, we observe cascade communications that appear to be algorithmically created.

The conversation in the accounts used by the Chinese information operations is more complex primarily due to the fact that these accounts seem to be recently purchased by the Chinese government or government proxy, and the earlier histories of these accounts is varied. We observe that, even though Arabic and Portuguese have a large proportion of the conversation by volume, their use is relegated to a few accounts that are structurally segregated from the rest of the network. The Chinese and English language campaigns are much more intertwined as China directs their information campaign at the Western world and at Hong Kong. While the messaging is aimed at Hong Kong, it is not necessarily aimed internal to China since Twitter is blocked by China's firewall.

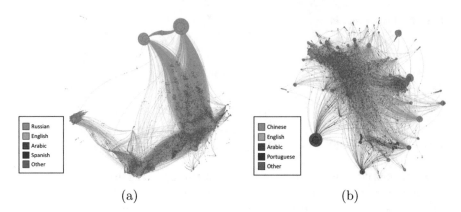

Fig. 1 The conversational network of the core accounts suspended and released by Twitter (colored by most recent language used by account). (**a**) Russia core conversation (**b**) China core conversation

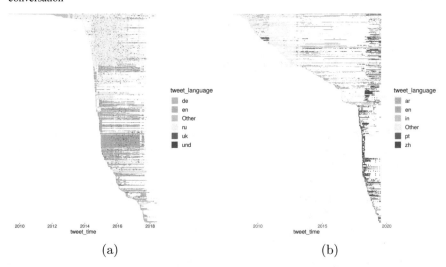

Fig. 2 Tweets over time colored by language. (**a**) Russia. (**b**) China

4.2 History of Accounts

In this section we will detail the history of these accounts. We believe that Fig. 2 produces a good backdrop to explaining each of these campaigns and the differences between them. Each row in this graph is a single account with it's tweets represented as points over time. This is colored by language (top 5 languages).

In the case of the Russian IRA, the timeline demonstrates a persistent effort to embed in both Russian and English language societies. Specific accounts embedded into target cultures and subcultures, learned to interact within the values and beliefs of the subculture, and then began to manipulate both the narrative and the network in these subcultures. We do see some evidence of dormancy with some accounts leaving the conversation for sometimes years at a time, but nonetheless consistent effort to curate virtual persona's within a narrow context.

In the case of the Chinese disinformation effort, we see a very different approach. These accounts use multiple languages, exhibiting that these personas are not consistently embedding in the same networks and conversations. We also see long dormancy periods where these accounts are likely waiting to be activated or sold to a new bot handler. Then suddenly they all appear to be acquired or otherwise activated and begin tweeting in Chinese. This narrative accounts for the wide variety of languages and topics that baffled the cursory data explorer.

The history of the accounts shows a very different approach between the two information campaigns. The Russian effort demonstrates a planned and persistent effort to embed into the target society, and especially within target subcultures. They did this in the Russian language to manipulate their own population, and in English to manipulate beliefs and actions in America. Once embedded these agents continued to develop a following and influence a larger and larger swath of the American populace.

The Chinese approach was much more reactionary, seems less planned, and did not have any persistent effort to embed in networks to affect influence. To some extent, the Chinese effort was simply to spam their narrative across the international community.

4.3 Geography of Accounts

While the geography of both of these data sets have been explored to some extent, we wanted to take a little different approach to the geography of Twitter data. In our analysis here, we focus on the national flags that are often added to an actor's description field in Twitter. These flag emoji's are produced by using ISO 3166-1 internationally recognized two-letter country codes. Examples of flag emoji's are shown here ▬ ❙❙ ☲ ❙❙ ▬ ❊. Flags are naturally used by individuals to associate themselves with a national identity. At times, individuals use multiple national flags in their description. Multiple national identities may be the result of immigration or a proud ex-patriot.

In our analysis of disinformation streams, however, we've seen bots and other malicious accounts use two or more flags in their profile. We believe that this is done so that an actor can leverage a curated and popular account in multiple target audiences and conversations. In particular we've seen this done with accounts so that they can participate in political conversations in North America and Europe, possibly in different languages, and make it looks as if they're just a passionate ex-patriot.

We found evidence of this behavior in the core data set, particularly in the IRA data. Two examples are ▬ in ▤ and Russian ▬ living in the US▤. In these cases, a description like this allows the casual observer to rationalize why the account

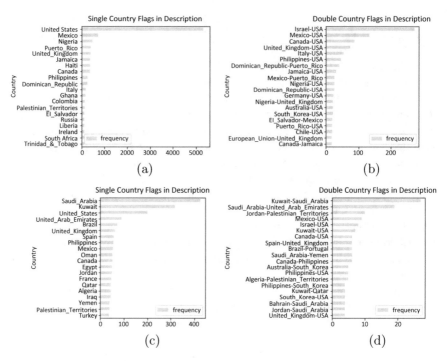

Fig. 3 The distribution of flag Emoji's in account descriptions. The high volume of unexpected flags used for the China data (such as Kuwait/Saudi Arabia) is due to the fact that many of these accounts were recently purchased by the Chinese government, and therefore most tweets and account descriptions by these accounts were produced by their previous owners. (**a**) IRA single flag. (**b**) IRA double flag. (**c**) China single flag. (**d**) China double flag

switches back and forth between Russian and English and between Russian social/political conversations and American social/political issues.

To explore this at scale, we developed algorithms that would extract the flag emoji's and build distributions. When we did this we built a distribution of single occurring flags and then of multiple flag combinations. The results of this analysis are provided in Fig. 3. In particular we see a high number of US-Israel flags combinations among the Russian information operations. Also of note is a high number of US-Italian combinations. While many of these may be legitimate, we have observed some accounts in different data that are simultaneously meddling in US political debate in English while encouraging Italy to leave the European Union in Italian.

4.4 Calculating Content Marketshare Over Time

Although we've already looked at the histories of these accounts, we wanted to understand temporal distributions better so that we can understand how these

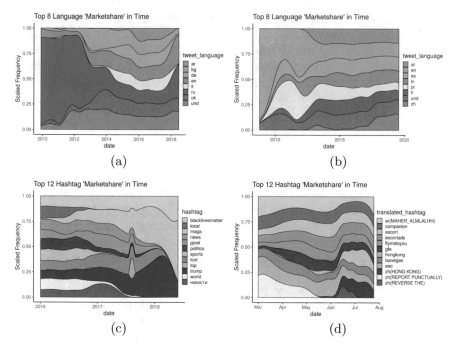

Fig. 4 Normalized marketshare of language and hashtags for core IRA and Chinese data suspended by Twitter. (**a**) IRA language market share. (**b**) China language market share. (**c**) IRA hashtag market share. (**d**) China hashtag market share

accounts were used over their life span as well as in the world events they're respectively associated with. To do this we explored the use of language and content over time with temporal market share.

To compute the temporal market share of language and hashtags we identified the top 8 languages and the top 12 hashtags in the core data for each operation, and their normalized portion (or market share) of the conversation over time. We see the visualization in Fig. 4. In the IRA data (graphs on left), we see a clear transition of information operations conducted in Russian to begin manipulation in Ukraine, English and other languages almost exclusively focused on Europe and the West.

In the plot of IRA hashtag market share, two things jump out. The first is the sudden outsized growth of IRA support of the #MAGA hashtag and the American right. The IRA did infiltrate the American left, but not to the same extent as the American right. The second and equally alarming observation is the long term and persistent use of the #blacklivesmatter hashtag as some of the IRA agents embedded into the African American subculture. The final but equally important observation we see here is that many of the hashtags are associated with a standard news organization. Multiple accounts in the data attempted to appear as a local news source or news aggregator in order to have the appearance of legitimacy.

From the Chinese core data, we see a wide variety of languages with only a small uptick in Chinese language at the end. Likewise the hashtag plot only has a small uptick in English and Chinese use of Hong Kong at the end. While Twitter associated all of the accounts with deliberate operations by the Chinese, the actual volume of data associated with the Hong Kong protests is limited compared to the total volume over the life of these accounts.

4.5   Bot Analysis

Social media bots are any account that has some level of action being automated by a computer. On Twitter tweeting, retweeting, replying, quoting, liking, following, and general searching can all be automated. In this section we leverage several bot detection tools to predict the number of accounts that appear to have automated behavior. Memes and bots are tools used to conduct information maneuvers in influence campaigns [6].

The models used below are two external models as well as two that were developed by our team. The first external model is the Debot model [8]. The Debot model is an unsupervised model that finds bots that are correlated using warped correlation. In other words, this model finds two or more accounts that are posting the same content at roughly the same time. The Debot team continually monitors parts of Twitter, and keeps a database of accounts that they've found to be correlated. In our search through the Russia and Twitter periphery data, we searched the Debot database to identify any of our accounts that have been found before. The second external model is the Botometer model (previously called the BotOrNot model) [10]. The Botometer model is a supervised machine learning model with well documented feature space. The Botometer Application Programming Interface (API) accepts a user ID or screen names as input, scrapes the Twitter API using the consumer provided keys on the server side, and then returns a score for content, friends, network, sentiment, temporal, user, and the universal score for the account. Given this method, Botometer scores are only available for accounts that are still active (i.e. not suspended, private, or otherwise shutdown). Due to the time required to scrape the timeline, in both of our data sets we randomly sampled 5,000 accounts for the Botometer model.

We've also listed scores for two models developed internally. The Bot-Hunter suite of tools provides supervised bot detection at several data granularities. Tier 1 conducts bot detection with a feature space developed from the basic tweet JSON data that is returned by the Twitter API [4]. This includes features extracted from the *user* object and the *tweet* object. Tier 2 performs bot detection using the users timeline (adding more content and temporal features), and Tier 3 uses the entire conversation around an account to predict the bot score [3]. Due to the computational cost of running Tier 3 (approximately 5 min per account), it is best for only a handful of accounts and was not used on these data sets. The Bot-Hunter Tier 1 models was run on all data, and the Tier2 was run on a random sample of

Table 3 Bot prediction for *core* and *periphery* data (% of total)

	Russia IRA		China (Hong Kong)	
	Core	Periphery	Core	Periphery
Accounts		697,296		204,920
Debot	**	1.07%	**	0.66%
Botometer	**	$9.1 \pm 0.7\%$	**	$28.5 \pm 1.3\%$
Bot-Hunter Tier 1		13.20%		8.68%
Bot-Hunter Tier 2	9.35%	$15.9 \pm 0.9\ \%$		$13.8 \pm 0.9\%$
Suspended/Closed	100%	4.30%	100%	0.30%

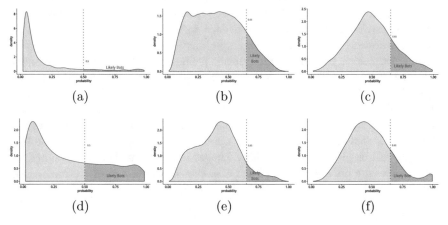

Fig. 5 Probability distributions for bot prediction for Botometer, Bot-Hunter(BH) Tier 1 and Tier 2 with threshold shown. (**a**) IRA botometer. (**b**) IRA BH-tier 1. (**c**) IRA BH-tier2. (**d**) China botometer. (**e**) China BH-tier 1. (**f**) China BH-tier 2

5000 accounts. Note that unlike Botometer, Bot-Hunter runs on existing data and was therefore able to predict on core, periphery, and suspended accounts. We've also developed an abridged version of Bot-Hunter Tier 1 that can run on the core data since it doesn't contain all features available for the unabridged model.

From Table 3 we see that models predict that 9–15% of the Russian core and periphery have likely automated behavior, with Hong Kong estimates slightly lower with Bot-Hunter predicting 8–14% automated behavior and Botometer as the outlier with 28% prediction.

We get even more insight into these models and data by looking at Fig. 5. This shows the probability distribution and chosen thresholds for each of the models on the periphery data. The biggest takeaway in these images is the difference between the shape of the Botometer model and the Bot-Hunter models. Although both are trained with a similar supervised learning model (Random Forest Classifier), they were trained on very different training data. Because of this, Botometer shows that most accounts are very unlike automated accounts, whereas Bothunter models show that the majority of accounts seem to appear a little more automated. Given that both models are similar, these distributions are saying that the these suspect accounts associated with Russian and Chinese disinformation are more similar to the data that Bot-Hunter was trained on than the data that Botometer was trained on.

4.6 Multi-media Analysis

Richard Dawkins originally created the word meme in his book Selfish Gene in which he defined a meme as a "...noun that conveys the idea of a unit of cultural transmission, or a unit of imitation" [11]. Shifman later adapted and defined internet memes as artifacts that "(a) share common characteristics of content, form, and/or stance; (b) are created with awareness of each other; and (c) are circulated, imitated, and transformed via the internet by multiple users" [20, 21].

Internet memes, particularly multi-media memes, are increasingly used in online information warfare. This phenomena has been highlighted in articles like the New York Time "The Mainstreaming of Political Memes Online" [7], and has been dubbed memetic warfare. To analyze memes in these two data sets, we developed a deep learning meme classifier to extract memes from the multi-media archives that Twitter shared along with the data. We ran this classifier on all images in the IRA data set, and on all Hong Kong related images in the China data set. Examples of IRA memes are provided in Fig. 6 and examples of China memes are provided in Fig. 7.

Fig. 6 Russian IRA memes

Fig. 7 China memes

From our analysis of these we see the IRA use memes at a much higher volume and sophistication. IRA memes involve significant creative content development and solid understanding of target subculture and biases. The IRA memes uses standard meme templates so that their memes fit in with the deluge of internet memes flowing around an election event. Their memes also cover the full spectrum of information operations forms of maneuver, to include all aspects of supporting or attacking narratives and supporting or attacking networks.

The Chinese memes, in contrast, were hastily created and are primarily informational in nature. In fact, in many respects they are do not meet the definition of a meme that Dawkins and Shifman put forward above, since we do not see significant evolution and transformation by multiple users. To some extent, they represent another facet of a campaign to spam the world with a particular narrative about the Hong Kong protests. Across their information campaigns, the Chinese seem reluctant to uses memes. While part of this may be cultural, another reason for their reluctance may be a worry that the evolution and propagation of memes is in the hands of the masses, not tightly controlled by central authorities. Memes can quickly turn negative toward the Chinese Communist Party and its leadership, as they did with Winnie the Pooh memes, causing party leadership to ban and censor Winnie the Pooh memes [15].

4.7 State Sponsored Accounts

Part of the role of bots within information operations is to amplify certain voices within the conversation. With state sponsored information operations, this often means amplifying state sponsored media. In recent years, Russia has increased the

Table 4 Add caption

State owned media	IRA core data	Chinese core data
Russian	72,846	8
Chinese	226	1,400
American	11	0
Korean	0	2
German	62	2

worldwide penetration of RT, Sputnik, and other state sponsored news agencies, while China has been gaining greater international penetration with China Xinhua News. To measure the extent that to which this data is amplifying these voices, we collected a large list of all Twitter handles associated with these Russian and Chinese state owned media companies, as well as handles associated with several other country's state owned media (for example the US Voice of America) for comparison. While the degree to which each of these handles spread state "propaganda" varies widely, we provide them for comparison.

We then scanned the *core* data for both datasets to examine the degree to which each data set amplifies these state owned media company's. The results are provided in Table 4.

As can be seen by this table, both the Chinese and especially the Russian dataset provides massive amplification for these state owned media.

5   How Many Similar Actors are Left?

One of the biggest questions that remains after going through this data is "How many state sponsored actors are still at large in the virtual world and currently manipulating world events?" To try to answer this question we spend some time analyzing the *periphery* data that is still mostly 'alive' and active on Twitter. Some of these actors may have been randomly brought into the data set, possibly by bots that were randomly mentioning normal citizens on Twitter in an effort to build a following/friend tie and begin to influence them. Others, however, are undoubtedly part of the larger information campaign and are still conducting malicious and divisive operations.

As shown above, at ~10% of both streams exhibit bot like behavior (these are again conservative estimates). Of the accounts in the periphery, 85.2% of the Russian accounts and 64% of the Chinese accounts are active, meaning they are not dormant and have tweeted in the last 6 months. Additionally, these accounts continue to amplify state owned propaganda. The IRA *periphery* amplifies Russian state owned media 6,023 times, and the China *periphery* amplifies Chinese state owned media 1,641 times.

Below we try to capture the primary topics that these accounts are embedding in. To do this we sampled 5000 accounts from the periphery of Russia and from China,

$$\text{(a)} \qquad\qquad\qquad\qquad \text{(b)}$$

Fig. 8 Current Information Operations by Russia found in the *periphery* data. (**a**) Possible Russia influence in the US Left (8–10% of Accounts). (**b**) Possible Russia influence in the US Right (10–12% of Accounts)

collected the last 200 tweets associated with each accounts. After selecting only those tweets in the last 6 montsh, we conducted topic analysis with latent dirichlet allocation (LDA). By optimizing the Calinski Harabaz score, we chose a k of 10 for LDA.

The Russian data shows clear topic groups that are attempting to meddle in Western affairs. The wordclouds of two of these topic groups is shown in Fig. 8. These images show a continued effort to divide America by further polarizing an already polarized political climate. Note that other topics not shown here include efforts to meddle in Europe (particularly amplifying the voice of the Yellow Vest Movement as well as far right groups), meddle in Canadian elections (clearly seen in the prominent place of #cdnpoli and #elxn43 in one LDA topic group of every sample tested).

From this we find that the Chinese data is still too diverse. The periphery data is associated with the entire timeline of these accounts, and is therefore too diverse to define clear information operation efforts and identify them in topics. During LDA and further analysis we found ~190 K accounts associated with Hong Kong, but they seemed to be across the spectrum of the discussion without any strongly coordinated disinformation operations (at least not in this periphery data). With the LDA analysis, we did find one sizable group that appeared to be against the current US administration. Once again, because of the randomness of the data it was difficult to claim this was due to a coordinated effort and not just caused by random bot behavior.

6   Conclusion

Throughout the data we see an experienced, sophisticated and well resourced campaign by Russia's Internet Research Agency while we also observe a Chinese campaign that appears reactionary and ad hoc. Several major conclusions are summarized below:

- The IRA's effort included identification and study of target subcultures with significant effort to shape messaging to leverage existing biases.
- The Chinese effort was aimed at Hong Kong and the international community at large without evidence of extensive effort to identify or a target audience or craft messaging for a specific audience.
- The IRA effort demonstrates an understanding of internet memes and a willingness to take risks in releasing multi-media messaging that will evolve in the masses.
- The Chinese effort demonstrates an unwillingness to release internet memes that will evolve outside of the direct control of central authorities.
- Both efforts, but particularly the Russian effort, demonstrate an effort to use these covert information operations to enhance the overt information operations conducted by state owned media companies.

While the focus of this research is on manipulation by well resourced nation-states, these same tactics can and are being used by smaller nation states (Saudi Arabia, Iran, Venezuela) and by non-state actors such as ISIS.

This work lays the foundation for building a repeatable end-to-end process for characterizing malicious actors in disinformation streams in social media, which is essential for national security [5]. These efforts to characterize actors will assist social cybersecurity analysts and researchers in getting beyond the binary classification of 'bot or not.' Future research will describe and illustrate this full workflow and several different data sets.

Acknowledgements This work was supported in part by the Office of Naval Research (ONR) Award N00014182106 and Award N000141812108, and the Center for Computational Analysis of Social and Organization Systems (CASOS). The views and conclusions contained in this document are those of the authors and should not be interpreted as representing the official policies, either expressed or implied, of the ONR or the U.S. government.

References

1. Alvarez, P., Hosking, T.: The full text of Mueller's indictment of 13 Russians'. The Atlantic, 16th Feb (2018)
2. Badawy, A., Addawood, A., Lerman, K., Ferrara, E.: Characterizing the 2016 Russian IRA influence campaign. Soc. Netw. Anal. Min. **9**(1), 31 (2019)
3. Beskow, D., Carley, K.M.: Bot conversations are different: leveraging network metrics for bot detection in twitter. In: Advances in Social Networks Analysis and Mining (ASONAM), 2018 International Conference on, pp. 176–183. IEEE (2018)
4. Beskow, D., Carley, K.M.: Introducing bothunter: a tiered approach to detection and characterizing automated activity on twitter. In: Bisgin, H., Hyder, A., Dancy, C., Thomson, R. (eds.) International Conference on Social Computing, Behavioral-Cultural Modeling and Prediction and Behavior Representation in Modeling and Simulation. Springer (2018)
5. Beskow, D.M., Carley, K.M.: Army must regain initiative in social cyberwar. Army Mag. **69**(8), 24–28 (2019)

6. Beskow, D.M., Carley, K.M.: Social cybersecurity: an emerging national security requirement. Mil. Rev. **99**(2), 117 (2019)
7. Bowles, N.: The mainstreaming of political memes online. New York Times (Feb 2018). https://www.nytimes.com/interactive/2018/02/09/technology/political-memes-go-mainstream.html
8. Chavoshi, N., Hamooni, H., Mueen, A.: Debot: twitter bot detection via warped correlation. In: ICDM, pp. 817–822 (2016)
9. Chen, A.: The agency. N. Y. Times **2**(6), 2015 (2015)
10. Davis, C.A., Varol, O., Ferrara, E., Flammini, A., Menczer, F.: Botornot: a system to evaluate social bots. In: Proceedings of the 25th International Conference Companion on World Wide Web, pp. 273–274. International World Wide Web Conferences Steering Committee (2016)
11. Dawkins, R.: The Selfish Gene: With a New Introduction by the Author. Oxford University Press, Oxford (2006). (Originally published in 1976)
12. DiResta, R., Shaffer, K., Ruppel, B., Sullivan, D., Matney, R., Fox, R., Albright, J., Johnson, B.: The Tactics & Tropes of the Internet Research Agency. New Knowledge, New York (2018)
13. Faris, R., Villeneuve, N.: Measuring global internet filtering. In: Access Denied: The Practice and Policy of Global Internet Filtering, vol. 5. MIT Press, Cambridge (2008)
14. Howard, P.N., Ganesh, B., Liotsiou, D., Kelly, J., François, C.: The IRA, Social Media and Political Polarization in the United States, 2012–2018. University of Oxford, Oxford (2018)
15. McDonell, S.: Why China censors banned Winnie the pooh – BBC news. https://www.bbc.com/news/blogs-china-blog-40627855 (July 2017). Accessed 29 Sep 2019
16. Mueller, R.S.: Report on the Investigation into Russian Interference in the 2016 Presidential Election. US Department of Justice, Washington (2019)
17. Nimmo, B., Brookie, G., Karan, K.: #trolltracker: twitter troll farm archives – DFRLAB – medium. file:///Users/dbeskow/Dropbox/CMU/bot_labels/references/ira/%23TrollTracker_%20Twitter%20Troll%20Farm%20Archives%20-%20DFRLab%20-%20Medium.html. Accessed 23 Sep 2019
18. Safety, T.: Information operations directed at Hong Kong. https://blog.twitter.com/en_us/topics/company/2019/information_operations_directed_at_Hong_Kong.html (August 2019). Accessed 26 Sep 2019
19. National Academies of Sciences, Engineering, and Medicine: A Decadal Survey of the Social and Behavioral Sciences: A Research Agenda for Advancing Intelligence Analysis. The National Academies Press, Washington (2019). https://doi.org/10.17226/25335, https://www.nap.edu/catalog/25335/a-decadal-survey-of-the-social-and-behavioral-sciences-a
20. Shifman, L.: The cultural logic of photo-based meme genres. J. Vis. Cult. **13**(3), 340–358 (2014)
21. Shifman, L.: Memes in digital culture. MIT Press, London (2014)
22. Uren, T., Thomas, E., Wallis, J.: Tweeting Through the Great Firewall: Preliminary Analysis of PRC-linked Information Operations on the Hong Kong Portest. Australia Strategic Policy Institute: International Cyber Policy Center, Barton (2019)
23. Von Nordheim, G., Boczek, K., Koppers, L.: Sourcing the sources: An analysis of the use of twitter and facebook as a journalistic source over 10 years in the New York times, the guardian, and süddeutsche zeitung. Digit. Journal. **6**(7), 807–828 (2018)

Pretending Positive, Pushing False: Comparing Captain Marvel Misinformation Campaigns

Matthew Babcock, Ramon Villa-Cox, and Kathleen M. Carley

Abstract It has become apparent that high-profile Twitter conversations are polluted by individuals and organizations spreading misinformation on the social media site. In less directly political or emergency-related conversations there remain questions about how organized misinformation campaigns occur and are responded to. To inform work in this area, we present a case study that compares how different versions of a misinformation campaign originated and spread its message related to the Marvel Studios' movie Captain Marvel. We found that the misinformation campaign that more indirectly promoted its thesis by obscuring it in a positive campaign, #AlitaChallenge, was more widely shared and more widely responded to. Through the use of both Twitter topic group analysis and Twitter-Youtube networks, we also show that all of the campaigns originated in similar communities. This informs future work focused on the cross-platform and cross-network nature of these conversations with an eye toward how that may improve our ability to classify the intent and effect of various campaigns.

Keywords Misinformation · Network analysis · Social media · Topic groups · Cross-platform

1 Introduction

Inaccurate and misleading information on social media can be spread by a variety of actors and for a variety of purposes. In potentially more immediately higher stakes Twitter conversations such as those focused on natural disasters or national elections, it is taken for granted that pressure groups, news agencies, and other larger and well-coordinated organizations or teams are part of the conversation and may spread misinformation. In less directly political or emergency-related conversations,

M. Babcock (✉) · R. Villa-Cox · K. M. Carley
School of Computer Science, Carnegie Mellon University, Pittsburgh, PA, USA
e-mail: rvillaco@andrew.cmu.edu; kathleen.carley@cs.cmu.edu

© Springer Nature Switzerland AG 2020
K. Shu et al. (eds.), *Disinformation, Misinformation, and Fake News in Social Media*,
Lecture Notes in Social Networks, https://doi.org/10.1007/978-3-030-42699-6_5

there remain questions of how such organized misinformation campaigns occur and how they shape discourse. For example, the Marvel Cinematic Universe comic book movies are very popular and the most financially successful film franchise to date. As the Marvel movies have expanded their casts and focused on more diverse storytelling and storytellers, they have become flashpoints in the United States for online expressions of underlying cultural debates (both those debated in good faith and those that are not). This trend appears to be part of a wider one in the comic book communities on social media and within comics' news and politics sites [20].

Social media platforms can be places where honest cultural and political debates occur. However, because of the presence of misinformation and propaganda, many such conversations are derailed, warped, or otherwise "polluted". Conversations that tend to polarize the community often carry misinformation [21], and such polluted content serves the purpose of the influencer or the creator of misinformation. Therefore, further investigation into the use of misinformation on Twitter and the groups that push it in the service of cultural debates may be useful in informing community, individual, and government responses for preserving healthy discussion and debate online. To inform future work in this area, we conducted a case study to explore a set of these issues as they appear in the Captain Marvel movie Twitter discussion.

Captain Marvel, Marvel Studio's first female lead-focused superhero movie, was released on March 8, 2019. Much of the Twitter discussion of the movie was standard comic book movie corporate and fan material. However, in the runup to the release of the movie there was also a significant amount of contentious discussion centered on whether the movie should be supported and/or whether illegitimate efforts were being made to support or attack it. The underlying core of the contentious discussions was related to recurring debates over diversity and inclusion in mass media in general and in comics in particular [20]. There were several different types of false or misleading information that was shared as part of these discussions. One of the most prominent examples was the claim that the lead actor, Brie Larson, had said that she did not want white males to see the movie, which was false. This and similar claims about Larson and Marvel were promoted in multiple ways and were the basis for campaigns calling for a boycott of the movie, either directly (#BoycottCaptainMarvel) or indirectly (by promoting an alternative female-led action movie, Alita: Battle Angel through a hijacked hashtag, #AlitaChallenge).[1]

As these campaigns were found to have somewhat similar community foundations but experienced differing levels of success, they make for a good case

---

[1]#AlitaChallenge first appeared on Twitter to support interest in the Alita movie a year before both the Alita: Battle Angel (Feb 2019) and Captain Marvel (March 2019) movies were released. Up and through the release of Alita the hashtag had minimal use. On March 4, 2019 a politically right-wing celebrity/conspiracy theorist used the hashtag to promote their version of the Alita Challenge (i.e. go spend money on Alita instead of Captain Marvel during the latter's opening weekend) and the hashtag soared in popularity. As Twitter users pointed out, this appears to have been disingenuous of the hijacker as prior tweets demonstrate their lack of interest in Alita as a movie.

study to explore the impact of organization and narrative framing on the spread of misinformation. Moreover, unlike what was observed on previous work on false information in Twitter comic book movie discussions [2], the sources of misinformation in this campaign were more well-known organized groups and/or more established communities providing an additional avenue of exploration. In this work, we compare the two misinformation-fueled boycott campaigns through examination of their origins, the actors involved, and their diffusion over time.

2   Related Work

The study of the spread of misinformation on platforms such as Twitter is a large and growing area of work that can be seen as part of a new emphasis on social cyber security [7, 11]. For example, much of that research has been concerned with who is spreading disinformation, and pointing to the role of state actors [18, 19], trolls [23] and bots [13, 27]. Other research has focused on how the media is manipulated [3, 17]. Still others have focused on impact of spreading disinformation on social processes such as democracy [4], group evolution [3], crowd manipulation [15], and political polarization [26]. There are also efforts aimed at collecting these different types of information maneuvers (e.g. bridging groups, distracting from the main point, etc) that were employed by misinformation campaigns into !arger frameworks [6].

Most of the research on information diffusion in social media has focused on retweeting. However, there are many reasons why people retweet [16] and retweeting is only one potential reaction to information found on Twitter. Examining simple retweet totals is static, and doing so does not provide insight into the entire diffusion path [1]. Identifying the full temporal path of information diffusion in social media is complex but necessary for a fuller understanding [22]. Moreover, people often use other mechanisms such as quoting to resend messages [8]. In doing so, they can change the context of the original tweet, as when quotes are used to call out and attack the bad behavior of the original tweeter [2]. Examining replies to tweets and the support of those replies also assists in creating a more complete picture of the discourse path. In contrast to earlier approaches, we consider the full temporal diffusion of misinformation and responses, and the roles of diverse types of senders including celebrities, news agencies and bots.

Many characteristics of message content may impact information diffusion [12]. For example, moral-emotional language leads messages to diffuse more within groups of like-minded individuals [9]. Emotionally charged messages are more likely to be retweeted [24] as are those that use hashtags and urls [25]. The use of urls shows the cross-platform nature of misinformation on social media with many urls pointing to other sites such as YouTube. Research only focusing on one ecosystem may mischaracterize both the who and how of misinformation spread. While our focus is on Twitter activity in the present work, we also use Twitter-Youtube connections to better understand the communities involved.

3 Data Description and Methods

From February 15 to March 15, 2019, we used Twitter's API to collect tweets for our analysis. Our goals and methods for tweet collection were as follows:

1. Compare the origination, spread, and response to the two main campaigns to push the misinformation campaign on interest. To do this we collected all non-reply/non-retweet origin tweets that used #BoycottCaptainMarvel and #AlitaChallenge during our period of interest and collect all quotes, replies and retweets of these origins.
2. Explore the conversation around these two campaigns that did not use the two main hashtags. Compare the hashtag-based and non-hashtag-based conversation. To do this we collected all non-reply/non-retweet origin tweets that used "Alita" along with one of a set of keywords used in the contentious comic-book Twitter discussions (e.g. "SJW", "Feminazi") and collect all quotes, replies and retweets of these origins. We labeled this as the "Charged Alita" conversation.
3. Characterize which communities users who spread or responded to the misinformation were from. To do this we collected the Twitter timelines of the central users and all non-reply/non-retweet origin tweets that provide information about the general Captain Marvel movie conversation (using the keywords #CaptainMarvel, Captain Marvel, Brie Larson), and all quotes, replies and retweets of these origins. To use cross-platform information to understand community structure we collected author, subject, and viewership information for all YouTube video URLs shared through Twitter.

For goals 1 and 3 we are relatively confident that our method allowed us to collect that vast majority if not the entirety of the tweets we aimed for due to our intentions to look at very specific hashtags and to obtain a general sense of where such hashtags were used compared to the most general conversation. For goal 2, while we able to collect a large enough sample of tweets related to the campaign, it is probable that some discussions of the Alita Challenge took place on Twitter using keywords we did not search for, and therefore are not part of our analysis.

We rehydrated any available target of a reply that was not originally captured in the first collection. This allowed us to capture at least the first level interaction within the relevant conversations. In total, we collected approximately 11 million tweets. We used a CASOS developed machine-learning tool [14] to classify the twitter users in our data as celebrities, news agencies, company accounts, and regular users. We used CMU BotHunter [5] to assign bot-scores to each account in our data set. We used the python wordcloud library to generate word clouds for the different groups we found based on YouTube video names.

We used ORA [10], a dynamic network analysis tool, to create and visualize Topic Groups and Louvain clustering that was used to explore the Twitter communities that shared and responded to the misinformation campaigns investigated. Topic Groups are constructed by using Louvain clustering on the intersection network between the Twitter user x Twitter user (all communication) network and the Twitter user x concept network. The resulting Topic Groups thus provide an estimation of which Twitter users in a data set are communicating with each other about the same issues.

4   Results

4.1   *Diffusion on Twitter*

We compared the diffusion of the original boycott campaign tweets and responses to them over time. As shown in Fig. 1, direct calls to boycott the Captain Marvel movie started more than a month before its official release on March 8, 2019, without gaining much traction on Twitter except for a day or two before the movie's release (green line). In contrast, during the same period there was an increase of discussion of "Alita" using harsh "culture war" phrases aimed at the *Captain Marvel* movie (Charged Alita). Most striking was the relatively rapid spread of #AlitaChallenge after March 4 (i.e. after the new use of it to attack Captain Marvel promoted by politically right-wing commentators). Overall, most support for these campaigns occurred between March 4 and March 12. Figure 1 also shows that most of the support for responses to the original campaign tweets occurred after the movie was released and based on visual inspection the majority of this support was for responses critical of the various boycott campaigns. The #AlitaChallenge campaign, while being the most directly supported, was also the most widely criticized, followed by responses to the Charged Alita conversations. It is noteworthy that the spike in negative responses coincided with the end of the majority of the support

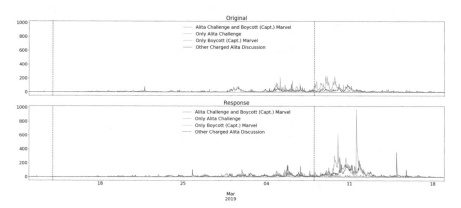

Fig. 1 Diffusion of boycott/anti-Captain Marvel Alita campaigns. Retweets per 30 min window of originals (top) and responses (bottom) for tweets that shared (1) both #AlitaChallenge and #BoycottCaptainMarvel, (2) only #AlitaChallenge, (3) only #BoycottCaptainMarvel, and (4) tweets with "Alita" and harsh words but no #AlitaChallenge. Vertical dashed lines are the opening of Alita: Battle Angel (leftmost) and Captain Marvel (rightmost)

for #AlitaChallenge or Charged Alita tweets. This behavior has been observed on other similar events with contentious messages and debunked rumors [2]. Tweets that directly pushed #BoycottCaptainMarvel or pushed both sets of hashtags were not responded to as great an extent.

4.2 Originating and Responding Twitter Communities

In order to explore which Twitter communities in the overall Captain Marvel conversation the #AlitaChallenge and #BoycottCaptainMarvel campaigns originated and spread to, we calculated the Topic Groups and constructed the Topic Group x Twitter user network. Prior to calculating the Topic Groups, we removed both main hashtags so that the resulting groups would not be based on them as inputs.

Figure 2 shows that a significant majority of origin tweets came from users who were found to be in topic group 8 or 13 and that origin tweets for all four campaign types were found in the same topic groups. Topic Group 8 includes the account that original highjacked #AlitaChallenge as well as accounts and concepts associated with Comicsgate controversies. Topic Group 13 appears to a group more focused on comic book movies in general. Figure 2 also shows that the responses to the different boycott campaigns were more spread out among several more topic groups.

In addition to using twitter actions to communicate (quotes, replies, and retweets), URLs were also shared and helpful in exploring whether the four campaigns were being shared in the same communities. Using the YouTube data mentioned above, we constructed an Agent x Agent network where the nodes are

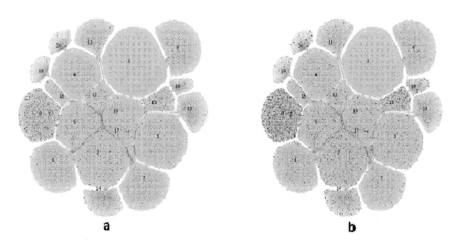

a   b

Fig. 2 Twitter Topic Groups Network. Numbers and light blue squares represent Topic Groups. Colored round nodes in Fig. 2a represent origin tweets of the four different types. Colored nodes in Fig. 2b represent the origin tweets and retweets of origin tweets in blue and replies and retweets of replies in green

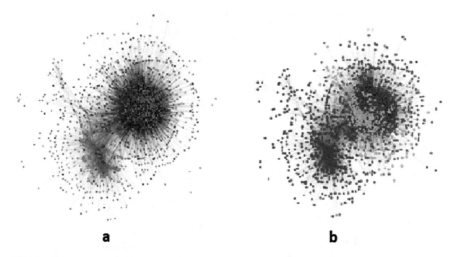

Fig. 3 Twitter user x Twitter user shared Youtube author network. Link weights at least 2. In Fig. 2a, the four colors represent the four types of boycott-related campaigns. Figure 2b is colored by louvain grouping

Twitter user accounts and the links represent the number of shared YouTube authors. For example, if Twitter user A tweets out YouTube videos by author X and Y over the course of our collection period and Twitter user B tweets out videos by X but not Y then the link between A and B is 1 (because they both shared videos by author X). As links with weights of 1 may not be that indicative of broader community (as it could signal only one shared video in addition to one shared author), we examined the subset of this network were the minimum link weight was 2 shared authors as shown in Fig. 2a, b:

Figure 3 shows that there is one main dense cluster of users and one smaller dense cluster. The main cluster was found by inspection to be Twitter accounts mostly involved with anti-Captain Marvel, anti-Marvel, right-wing politics, and "Red Pill" and Comicsgate-like misogyny or anti-diveristy sentiment. The smaller cluster (which is a separate louvain group as shown in Fig. 3b) appears to be Twitter users interested in comic book and other movies without an expressed political/cultural bent. Figure 3a shows that all four boycott related campaigns had origins and support from connected users in the main cluster (note that this cluster appears to have exisited prior to the release of Captain Marvel). It should also be pointed out that the Twitter user who began the hijacking of #AlitaChallenge is not central to the larger cluster though they are connected to the central cluster.

To further help identify the sentiment shared by each sub-cluster, we created word clouds using the titles of the YouTube videos shared by members of those groups as input. The results are shown in Fig. 4. While the common use of directly movie related words is most apparent in this figure, it can also be seen that Groups 2–5 (which are in the main cluster of Fig. 3) share to various degrees phrases that are either more critical of Marvel/Captain Marvel, involve Alita more, or involve

Fig. 4 Word clouds based on YouTube video titles shared by each Louvain group

political/cultural discussion points and targets. This is especially true of Groups 4 and 5 which contain anti-Democrat phrases, pro-Republican phrases (especially in Group 4) and phrases related to other movies with contentious diversity discussions such as Star Wars (Group 4 and 5). This result reinforces our understanding of the main cluster as being where many users that fall on the right-wing/anti-diversity side of such debates are located.

4.3 Types of Actors

We also identified the roles the different tweeters played on each of the conversations. As Table 1 shows the vast majority of actors involved in these conversations were regular Twitter users, followed by news-like agencies and reports and celebrities. Using a conservative bot-score we found low levels of bot-like accounts with the highest level being in the news-like category. Bot-like accounts did not appear to favor one type of interaction over the other (i.e. the percentage of bot-like accounts out of all the accounts that retweeted was similar to those that replied or originated stories), nor did they favor one type of story. Similarly low levels of bot-like account activity was observed in an analysis of a similar event [2].

Table 1 Roles of Twitter users in the four boycott-related campaigns. A bot-score of 0.8 or greater was used to classify bot-like agents

	Only #Ali-taChallenge	Only #BoycottCap-tainMarvel	Both hashtags	Only charged Alita
# Total actors	29,331	6,255	2,167	21,132
By actions (% bot-like):				
Original poster (4%)	3%	7%	10%	21%
Retweeter of original (4%)	33%	61%	51%	40%
Replier of original (5%)	27%	21%	18%	27%
Retweeter of replier (3%)	37%	11%	22%	12%
By identity (% bot-like):				
Regular (4%)	95.5%	96.1%	97.4%	93.4%
News/Reporters (7%)	2.5%	2.3%	1.0%	2.7%
Celebrity (2%)	1.1%	0.8%	1.3%	1.7%
Company (3%)	0.6%	0.5%	0.2%	2.0%
Government (3%)	0.3%	0.3%	0.1%	0.1%
Sports (3%)	0.1%	0.0%	0.0%	0.1%

It should be noted that the #AlitaChallenge discussion while involving the most users, was originated by a smaller percentage of those users when compared to other campaigns. The #BoycottCaptainMarvel campaign had the highest level of actors who were involved in retweeting and spreading the origin tweets.

5   Discussion and Conclusions

We used the interesting case of the boycott-based-on-misinformation campaigns in the Captain Marvel twitter discussion to examine the origins, actors, and diffusion of misinformation. Overall, we found that the origins of all four types of campaigns were located in the same communities that share right-wing/anti-diversity sentiments through YouTube. We found that though the origins and discussion take place in similar communities and with a similar presence of celebrity and news-like accounts, the diffusion of the more indirect #AlitaChallenge and the more direct #BoycottCaptainMarvel occurred in different ways. Here we want to touch upon some of these differences, discuss why they may have come about, and where this research can lead.

#AlitaChallenge and related Charged Alita discussions traveled to a much greater extent than #BoycottCaptainMarvel. This may have occurred because the #AlitaChallenge campaign on the surface presented a negative activity (boycotting Captain Marvel based on misinformation) as a positive one (supporting the strong female lead in Alita). Positive framing of actions has been found in past research to motivate participation, though not always. Our results may more strongly support the idea that #AlitaChallenge was more successful due to (1) successful hijacking of the hashtag by someone connected to the right-wing/anti-diversity comic book movie community and (2) the coordinated retweeting of a group of similar attempts to push #AlitaChallenge. It should be noted that while the #AlitaChallenge and related campaigns spread more widely than #BoycottCaptainMarvel, they also garnered a much larger response that appears predominately negative. The success of #AlitaChallenge and origin from a specific set of known actors may have been a contributor in making pro-Captain Marvel users more aware and galvanized to respond. The lower level of the #BoycottCaptainMarvel campaign also aligned with a lower level of responses (again most of the users in this discussion were just retweeting the original false-message based campaign).

Our other contribution in this work is demonstrating the possible utility of using both topic groups and cross-platform connections to better define communities of interest. Examining the commonalities in YouTube authors among Twitter users in our data provided a clearer picture of the communities that started and spread misinformation related to the Captain Marvel lead actress. Further examination of the connections between shared YouTube content and Twitter discussions could be useful in looking at the other misinformation being spread about the movie (both from the pro-movie and anti-movie sides). Another future avenue of exploration is to focus on the role those users who cross Twitter topic group or YouTube author clusters play in spreading or responding to specific campaigns.

Additional limitations to this work that could be improved upon in the future include the lack of a more fine-grained picture of the responses to the boycott campaigns, the need for additional Twitter network to YouTube network comparisons, and deeper examination of the cultural context in which these campaigns were taking place.

The purpose of this work should not be taken to be to help in the design of successful misinformation but rather to assist in the understanding of different methods intentionally used to promote false information. This may help with the design of effective and efficient community level interventions – at a minimum as an example of the kinds of campaigns to be aware of and vigilant against. Future work should delve more deeply into the cross-platform and cross-network nature of these conversations with an eye toward how that may improve our ability to classify the intent and effect of various campaigns.

References

1. Amati, G., Angelini, S., Gambosi, G., Pasquin, D., Rossi, G., Vocca, P.: Twitter: temporal events analysis. In: Proceedings of the 4th EAI International Conference on Smart Objects and Technologies for Social Good. pp. 298–303. ACM (2018)
2. Babcock, M., Beskow, D.M., Carley, K.M.: Different faces of false: the spread and curtailment of false information in the black panther twitter discussion. J. Data Inf. Qual. (JDIQ) 11(4), 18 (2019)
3. Benigni, M.C., Joseph, K., Carley, K.M.: Online extremism and the communities that sustain it: detecting the isis supporting community on twitter. PLoS One 12(12), e0181405 (2017)

4. Bennett, W.L., Livingston, S.: The disinformation order: disruptive communication and the decline of democratic institutions. Eur. J. Commun. **33**(2), 122–139 (2018)
5. Beskow, D.M., Carley, K.M.: Bot-hunter: a tiered approach to detecting & characterizing automated activity on twitter. In: Conference paper. SBP-BRiMS: International Conference on Social Computing, Behavioral-Cultural Modeling and Prediction and Behavior Representation in Modeling and Simulation (2018)
6. Beskow, D.M., Carley, K.M.: Social cybersecurity: an emerging national security requirement. Mil. Rev. **99**(2), 117 (2019)
7. Bhatt, S., Wanchisen, B., Cooke, N., Carley, K., Medina, C.: Social and behavioral sciences research agenda for advancing intelligence analysis. In: Proceedings of the Human Factors and Ergonomics Society Annual Meeting, vol. 63, pp. 625–627. SAGE Publications: Los Angeles (2019)
8. Boyd, D., Golder, S., Lotan, G.: Tweet, tweet, retweet: conversational aspects of retweeting on twitter. In: 2010 43rd Hawaii International Conference on System Sciences, pp. 1–10. IEEE (2010)
9. Brady, W.J., Wills, J.A., Jost, J.T., Tucker, J.A., Van Bavel, J.J.: Emotion shapes the diffusion of moralized content in social networks. Proc. Natl. Acad. Sci. **114**(28), 7313–7318 (2017)
10. Carley, K.M.: Ora: a toolkit for dynamic network analysis and visualization. In: Encyclopedia of Social Network Analysis and Mining, pp. 1219–1228. Springer, New York (2014)
11. Carley, K.M., Cervone, G., Agarwal, N., Liu, H.: Social cyber-security. In: International Conference on Social Computing, Behavioral-Cultural Modeling and Prediction and Behavior Representation in Modeling and Simulation, pp. 389–394. Springer (2018)
12. Ferrara, E.: Manipulation and abuse on social media by Emilio Ferrara with Ching-man Au Yeung as coordinator. ACM SIGWEB Newsletter (Spring) **4**, 1–9 (2015)
13. Ferrara, E.: Measuring social spam and the effect of bots on information diffusion in social media. In: Complex Spreading Phenomena in Social Systems, pp. 229–255. Springer, Cham (2018)
14. Huang, B.: Learning User Latent Attributes on Social Media. Ph.D. thesis, Carnegie Mellon University (2019)
15. Hussain, M.N., Tokdemir, S., Agarwal, N., Al-Khateeb, S.: Analyzing disinformation and crowd manipulation tactics on youtube. In: 2018 IEEE/ACM International Conference on Advances in Social Networks Analysis and Mining (ASONAM), pp. 1092–1095. IEEE (2018)
16. Macskassy, S.A., Michelson, M.: Why do people retweet? anti-homophily wins the day! In: Fifth International AAAI Conference on Weblogs and Social Media (2011)
17. Marwick, A., Lewis, R.: Media Manipulation and Disinformation Online. Data & Society Research Institute, New York (2017)
18. Mejias, U.A., Vokuev, N.E.: Disinformation and the media: the case of Russia and Ukraine. Media Cult. Soc. **39**(7), 1027–1042 (2017)
19. Mueller, R.S.: The Mueller Report: Report on the Investigation into Russian Interference in the 2016 Presidential Election. WSBLD (2019)
20. Proctor, W., Kies, B.: On toxic fan practices and the new culture wars. Participations **15**(1), 127–142 (2018)
21. Ribeiro, M.H., Calais, P.H., Almeida, V.A., Meira Jr, W.: "everything i disagree with is# fakenews": correlating political polarization and spread of misinformation. arXiv preprint arXiv:1706.05924 (2017)
22. Scharl, A., Weichselbraun, A., Liu, W.: Tracking and modelling information diffusion across interactive online media. Int. J. Metadata Semant. Ontol. **2**(2), 135–145 (2007)
23. Snider, M.: Florida school shooting hoax: doctored tweets and Russian bots spread false news. USA Today (2018)
24. Stieglitz, S., Dang-Xuan, L.: Emotions and information diffusion in social media—sentiment of microblogs and sharing behavior. J. Manag. Inf. Syst. **29**(4), 217–248 (2013)
25. Suh, B., Hong, L., Pirolli, P., Chi, E.H.: Want to be retweeted? large scale analytics on factors impacting retweet in twitter network. In: 2010 IEEE Second International Conference on Social Computing, pp. 177–184. IEEE (2010)

26. Tucker, J.A., Guess, A., Barberá, P., Vaccari, C., Siegel, A., Sanovich, S., Stukal, D., Nyhan, B.: Social media, political polarization, and political disinformation: a review of the scientific literature. Political Polarization, and Political Disinformation: A Review of the Scientific Literature (March 19, 2018)
27. Uyheng, J., Carley, K.M.: Characterizing bot networks on twitter: an empirical analysis of contentious issues in the Asia-Pacific. In: International Conference on Social Computing, Behavioral-Cultural Modeling and Prediction and Behavior Representation in Modeling and Simulation, pp. 153–162. Springer (2019)

Bots, Elections, and Social Media: A Brief Overview

Emilio Ferrara

Abstract Bots, software-controlled accounts that operate on social media, have been used to manipulate and deceive. We studied the characteristics and activity of bots around major political events, including elections in various countries. In this chapter, we summarize our findings of bot operations in the context of the 2016 and 2018 US Presidential and Midterm elections and the 2017 French Presidential election.

Keywords Social media · Bots · Influence · Disinformation

1 Introduction

Social media have been widely portrayed as enablers of democracy [12, 15, 47, 48, 50]. In countries where freedom to communicate and organize are lacked, social media provided a platform to openly discuss political [2, 9, 13, 23, 25, 55, 87] and social issues [8, 18, 19, 37, 38, 77, 82], without fears for safety or retaliation. Such platforms have also been used to respond to crises and emergencies [34, 49, 75, 88, 89]. It is hard to overstate the importance of these platforms for the billions of people who use them every day, all over the world.

However, as it happens with most powerful emerging technologies, the rise of popularity led to abuse. Concerns about the possibility of manipulating public opinion using social media have been brought a decade before they materialized [39]. Ample evidence was provided by the scientific community that social media can influence people's behaviors [5, 14, 31, 32, 45, 60]. These concerns have been corroborated by numerous recent studies [26–28, 40, 58, 66, 68, 81].

Social media can be used to reach millions of people using targeted strategies aimed to maximize the spread of a message. If the goal is to manipulate public

---

E. Ferrara (✉)
USC Information Sciences Institute, Marina del Rey, CA, USA
e-mail: emiliofe@usc.edu

© Springer Nature Switzerland AG 2020
K. Shu et al. (eds.), *Disinformation, Misinformation, and Fake News in Social Media*,
Lecture Notes in Social Networks, https://doi.org/10.1007/978-3-030-42699-6_6

opinion, one way to achieve it is by means of bots, software-controlled social media accounts whose goal is to mimic the characteristics of human users, while operating at much higher pace at substantially no downside for their operators. Bots can emulate all basic human activity on social media platforms, and they become increasingly more sophisticated as new advancements in Artificial Intelligence emerge [30, 41, 57, 70, 80].

In this chapter, we focus on the use of bots to manipulate the political discourse. The first anecdotal accounts of attempts to steer public opinion on Twitter date back to the 2010 US Midterm election [65] and similarly during the 2010 US Senate special election in Massachusetts [58, 62], where bots were used to generate artificial support for some candidates and to smear their opponents.

Attribution, i.e., the determination of the actors behind such operations, has proven challenging in most such cases [30]. One notorious exception is represented by the attribution of an interference campaign occurred during the 2016 US Presidential election to a Russian-sponsored operation. This was as a result of a thorough investigation on Russian interference led by the US Senate Select Committee on Intelligence (SSCI). They found that "The Russian government interfered in the 2016 U.S. presidential election with the goal of harming the campaign of Hillary Clinton, boosting the candidacy of Donald Trump, and increasing political and social discord in the United States."[1] Numerous studies have investigated the events associated with this operation [7, 10, 44].

It is worth noting that bots have been used for other purposes, for example social spam and phishing [29, 42, 43, 61, 69, 78, 79, 85]. Albeit much work has been devoted to the challenges of detecting social spam [35, 56, 90] and spam bots [11, 51, 52, 61, 72], only recently the research community started to investigate the effects that bots have on society, political discourse, and democracy. The goal of this chapter is to summarize some of the most important results in this space.

1.1 Contributions of This Chapter

The aim of this chapter is to connect results of our investigations into three major political events: (i) the 2016 US Presidential election; (ii) the 2017 French Presidential election; and (iii) the 2018 US Midterm elections. We will discuss the role of bots in these events, and highlight the influence they had on the online political discourse. The contributions of this chapter are as follows:

- We first provide a brief overview of how bots operate and what are the challenges in detecting them. Several recent surveys have been published on the problem of characterizing and detecting bots [71, 86], including our own on *Communications of the ACM* [30].

---

[1] See Wikipedia: https://en.wikipedia.org/wiki/Russian_interference_in_the_2016_United_States_elections

- We then illustrate our first, and maybe the most prominent, use case of bots-driven interference in political discourse, discussing how bots have been used during the 2016 US Presidential election to manipulate the discussion of the presidential candidates. This overview is based on our results that appeared prior to the November 8, 2016 election events [10].
- We then illustrate how bots have been used to spread disinformation prior to the 2017 French Presidential election to smear Macron's public image.
- Finally, we overview recent results that suggest how bots have been evolving over the course of the last few years, focusing on the 2018 US Midterm elections, and we discuss the challenges associated to their detection.

2 Anatomy of a Bot

2.1 What Is a Bot

In this chapter, we define *bot* (short for *robot*, a.k.a., social bot, social media bot, social spam bot, or sybil account) as a social media account that is predominantly controlled by software rather than a human user. Although the definition above inherently states nothing about the intents behind creating and operating a bot, according to published literature, malicious applications of bots are reported significantly more frequently than legitimate usage [30, 71].

While in this chapter we will focus exclusively on bots that aim to manipulate the public discourse, it is worth nothing that some researchers have used bots for social good [4, 60], as illustrated by a recent taxonomy that explores the interplay between intent and characteristics of bots [71]. Next, we describe some techniques to create and detect bots.

2.2 How to Create a Bot

In the early days of online social media, in the late 2000s, creating a bot was not a simple task: a skilled programmer would need to sift through various platforms' documentation to create a software capable of automatically interfacing with the front-end or the back-end, and operate functions in a human-like manner.

These days, the landscape has completely changed: indeed, it has become increasingly simpler to deploy bots, so that, in some cases, no coding skills are required to setup accounts that perform simple automated activities: tech blogs often post tutorials and ready-to-go tools for this purposes. Various source codes for sophisticated social media bots can be found online as well, ready to be customized and optimized by the more technically-savvy users [44].

We recently inspected same of the readily-available Twitter bot-making tools and compiled a non-comprehensive list of capabilities they provide [10, 28].

Most of these bots can run within cloud services or infrastructures like *Amazon Web Services* (AWS) or Heroku, making it more difficult to block them when they violate the Terms of Service of the platform where they are deployed.

A very recent trend is that of providing Bot-As-A-Service (BaaS): Advanced conversational bots powered by sophisticated Artificial Intelligence are provided by companies like *ChatBots.io* that can be used to carry digital spam campaigns [29] and scale such operations by automatically engaging with online users.

Finally, the increasing sophistication of Artificial Intelligence (AI) models, in particular in the area of *neural-based natural language generation*, and the availability of large pre-trained models such as OpenAI's GPT-2 [64], makes it easy to programmatically generate text content. This can be used to program bots that produce genuine-looking short texts on platforms like Twitter, making it harder to distinguish between human and automated accounts [3].

2.3   How to Detect Bots

The detection of bots in online social media platform has proven a challenging task. For this reason, it has attracted a lot of attention from the computing research community. Even DARPA, the U.S. *Defense Advanced Research Projects Agency*, became interested and organized the 2016 DARPA Twitter Bot Detection [74], with University of Maryland, University of Southern California, and Indiana University topping the challenge, focused on detecting bots pushing anti vaccination campaigns. Large botnets have been identified on Twitter, from dormant [24, 24], to very active [1].

The literature on bot detection has become very extensive. We tried to summarize the most relevant approaches in a survey paper recently appeared on the *Communications of the ACM* [30]: In that review, we proposed a simple taxonomy to divide the bot detection approaches into three classes: (i) bot detection systems based on social network information; (ii) systems based on crowd-sourcing and leveraging human intelligence; (iii) machine learning methods based on the identification of highly-predictive features that discriminate between bots and humans. We refer the interested reader to that review for a deeper analysis of this problem [30]. Other recent surveys propose complementary or alternative taxonomies that are worth considering as well [20, 20, 71, 86].

As of today, there are a few publicly-available tools that allow to do bot detection and study social media manipulation, including (i) Botometer,[2] a popular bot detection tool developed at Indiana University [21], (ii) BotSlayer,[3] an application

---

[2] Botometer: https://botometer.iuni.iu.edu/

[3] BotSlayer: https://osome.iuni.iu.edu/tools/botslayer/

that helps track and detect potential manipulation of information spreading on Twitter, and (iii) the Bot Repository,[4] a centralized database to share annotated datasets of Twitter social bots.

In conclusion, several algorithms have been published to detect bots using sophisticated machine learning techniques including deep learning [46], anomaly detection [22, 36, 59], and time series analysis [16, 73].

3   Social Media Manipulation

Bots have been reportedly used to interfere in political discussions online, for example by creating the impression of an organic support behind certain political actors [58, 62, 65, 66]. However, the apparent support can be artificially generated by means of orchestrated campaigns with the help of bots. This strategy is commonly referred to as social media *astroturf* [66].

3.1   2016 US Presidential Election

Our analysis of social media campaigns during the 2016 US Presidential Election revealed the presence of social bots. We here summarize our findings first published in [10], discussing data collection, bot detection, and sentiment analysis.

Data Collection   We manually crafted a list of hashtags and keywords related to the 2016 US Presidential Election with 23 terms in total, including 5 terms specifically for the Republican Party nominee Donald Trump, 4 terms for the Democratic Party nominee Hillary Clinton, and the remainder terms relative to the four presidential debates. The complete list of search terms is reported in our paper [10]. By querying the Twitter Search API between September 16 and October 21, 2016, we collected a large dataset. After post-processing and cleaning procedures, we studied a corpus constituted by 20.7 million tweets posted by nearly 2.8 million distinct users.

Bot Detection   We used Botometer *v1* (the version available in 2016) to determine the likelihoood that the most active accounts in this dataset were controlled by humans or were otherwise bots. To label accounts as bots, we use the 50% threshold—which has proven effective in prior studies [21, 30]—an account was considered to be a bot if the bot score was above 0.5. Due to the Twitter API limitations, it would have been impossible to test all the 2.78 million accounts in short time. Therefore, we tested the top 50 thousand accounts ranked by activity volume, which account for roughly 2% of the entire population and yet are responsible for producing over 12.6 million tweets, which is about 60% of the total

---

[4]Bot Repository: https://botometer.iuni.iu.edu/bot-repository/

conversation. Of the top 50 thousand accounts, Botometer classified as likely bots a total of 7,183 users (nearly 15%), responsible for 2,330,252 tweets; 2,654 users were classified as undecided, because their scores did not significantly diverge from the classification threshold of 0.5; the rest—about 40 thousand users (responsible for just 10.3 million tweets, less than 50% of the total)—were labeled as humans. Additional statistics are summarized in our paper [10].

Sentiment Analysis We leveraged sentiment analysis to quantify how bots (resp., humans) discussed the candidates. We used SentiStrength [76] to derive the sentiment scores of each tweet in our dataset. This toolkit is especially optimized to infer sentiment in short informal texts, thus ideally suited for social media. We tested it extensively in prior studies on the effect of sentiment on tweets' diffusion [32, 33]. The algorithm assigns to each tweet t a positive $P^+(t)$ and negative $P^-(t)$ polarity score, both ranging between 1 (neutral) and 5 (strongly positive/negative). Starting from the polarity scores, we captured the emotional dimension of each tweet t with one single measure, the sentiment score $S(t)$, defined as the difference between positive and negative polarity scores: $S(t) = P^+(t) - P^-(t)$. The above-defined score ranges between -4 and $+4$. The negative extreme indicates a strongly negative tweet, and occurs when $P^+(t) = 1$ and $P^-(t) = 5$. Vice-versa, the positive extreme identifies a strongly positive tweet labeled with $P^+(t) = 5$ and $P^-(t) = 1$. In the case $P^+(t) = P^-(t)$—positive and negative sentiment scores for a tweet t are the same—the sentiment $S(t) = 0$ of tweet t is considered neutral as the polarities cancel each other out.

Partisanship and Supporting Activity We used a simple heuristic based on the 5 Trump-supporting hashtags and the 4 Clinton-supporting to attribute user partisanships. For each user, we calculated their top 10 most used hashtags: If the majority supported one particular candidate, we assigned the given user to that political group (Clinton or Trump supporter). Compared to network-based techniques [6, 17], this simple partisanship assignment yielded a smaller yet higher-confidence annotated dataset, constituted by 7,112 Clinton supporters (590 bots and 6,522 humans) and 17,202 Trump supporters (1,867 bots and 15,335 humans).

Summary of Results: Engagement Figures 1 and 2 illustrate the Complementary Cumulative Distribution Functions (CCDFs) of replies and retweets initiated by bots

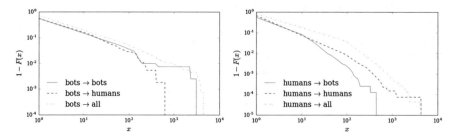

Fig. 1 Complementary cumulative distribution function (CCDF) of replies interactions generated by bots (left) and humans (right) (from [10])

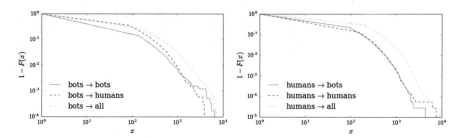

Fig. 2 Complementary cumulative distribution function (CCDF) of retweets interactions generated by bots (left) and humans (right) (from [10])

and humans in three categories: (i) within group (for example bot-bot, or human-human); (ii) across groups (e.g., bot-human, or human-bot); and, (iii) total (i.e., bot-all and human-all). The heavy-tailed distributions, typically observed in social systems, appear in both. Hence, further inspection of Fig. 1 suggests that (i) humans replied significantly more to other humans than to bots and, (ii) conversely, bots receive replies from other bots significantly more than from humans. One hypothesis is that unsophisticated bots could not produce engaging-enough questions to foster meaningful exchanges with humans.

Figure 2, however, demonstrates that retweets were a much more vulnerable mode of information diffusion: there is no statistically significant difference in the amount of retweets that humans generated by resharing content produced by other humans or by bots. In fact, humans and bots retweeted each other substantially at the same rate. This suggests that bots were very effective at getting their messages reshared in the human communication channels.

Our study highlighted a vulnerability in the information ecosystem at that time, namely that content was reshared often without a thorough scrutiny on the information source. Several subsequent studies hypothesized that bots may have played a role in the spread of false news and unverified rumors [67, 83].

Summary of Results: Sentiment We further explored how bots and humans talked about the two presidential candidates. Next, we show the sentiment analysis results based on *SentiStrength*. Figure 3 illustrates four settings: the top (resp., bottom) two panels show the sentiment of the tweets produced by the bots (resp., humans). Furthermore, the two left (resp., right) panels show the support for Clinton (resp., Trump). The main histograms in each panel show the volume of tweets about Clinton or Trump, separately, whereas the insets show the difference between the two. By contrasting the left and right panels we note that the tweets mentioning Trump are significantly more positive than those mentioning Clinton, regardless of whether the source is human or bot. However, bots tweeting about Trump generated almost no negative tweets and indeed produced the most positive set of tweets in the entire dataset (about 200,000 or nearly two-third of the total).

The fact that bots produce systematically more positive content in support of a candidate can bias the perception of the individuals exposed to it, suggesting that

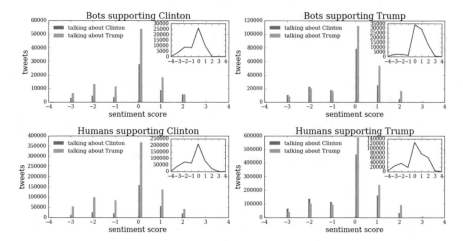

Fig. 3 Distributions of the sentiment of bots (top) and humans (bottom) supporting the two presidential candidates. The main histograms show the disaggregated volumes of tweets talking about the two candidates separately, while the insets show the absolute value of the difference between them (from [10])

there exists an organic, grassroots support for a given candidate, while in reality it is in part artificially inflated. Our paper reports various examples of tweets generated by bots, and the candidate they support [10].

3.2 2017 French Presidential Election

A subsequent analysis of the Twitter ecosystem highlighted the presence and effects of bots prior to the 2017 French Presidential Election. We next report our findings summarizing the results published in 2017 [28]. We provide a characterization of both the bots and the users who engaged with them.

Data Collection By following the same strategy as in the 2016 US Presidential election [10], we manually selected a set of hashtags and keywords related to the 2017 French Presidential Election. By construction, the list contained a roughly equal number of terms associated with each of the two candidates, namely Marine Le Pen and Emmanuel Macron, and various general election-related terms: we ultimately identified 23 terms, listed in our paper [28]. We collected data by using the Twitter Search API, from April 27 to the end of election day, on May 7, 2017: This procedure yielded a dataset containing approximately 17 million unique tweets, posted by 2,068,728 million unique users. Part of this corpus is a subset of tweets associated with the *MacronLeaks* disinformation campaign, whose details are described in our paper [28]. The timeline of the volume of posted tweets is illustrated in Fig. 4.

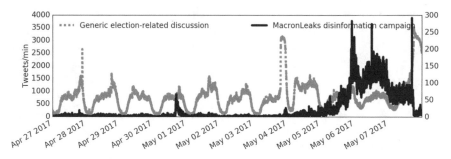

Fig. 4 Timeline of the volume of tweets generated every minute during our observation period (April 27 through May 7, 2017). The purple solid line (right axis) shows the volume associated with MacronLeaks, while the dashed grey line (left axis) shows the volume of generic election-related discussion. The presidential election occurred on May 7, 2017 (from [28])

Bot Detection Due to the limitations of the Twitter API, and the time restrictions for this short period of unfolding events, we were unable to run in real time the bot detection relying upon Botometer. For this reason, we carried out a post-hoc bot detection on the dataset using an offline version of the bot-detection algorithm inspired by Botometer's rationale. Specifically, we exclusively leveraged user metadata and activity features to create a simple yet effective bot detection classifier, trained on same data as Botometer, which is detailed in our paper [28]. We validated its classification accuracy and assessed that it was similar to Botometer's performance, with above 80% in both accuracy and AUC-ROC scores. Manual validation corroborated the performance analysis. Hence, we used this simplified bot detection strategy to unveil bots in the dataset at hand.

Summary: Temporal Dynamics We started by exploring the timeline of the general election-related discussion on Twitter. The broader discussion that we collected concerns the two candidates, Marine Le Pen and Emmanuel Macron, and spans the period from April 27 to May 7, 2017, the Presidential Election Day, see Fig. 4. Let us discuss first the dashed grey line (left axis): this shows the volume of generic election-related discussion. The discussion exhibits common circadian activity patterns and a slightly upwards trend in proximity to Election Day, and spikes in response to an off-line event, namely the televised political debate that saw Le Pen facing Macron. Otherwise, the number of tweets per minute averages between 300 and 1,500 during the day, and quickly approaches de facto zero overnight, consistently throughout the entire observation window. Figure 4 also illustrates with the purple solid line (right axis) the volume associated with MacronLeaks, the disinformation campaign that was orchestrated to smear Macron's reputation. The temporal pattern of this campaign is substantially different from the general conversation. First, the campaign is substantially silent for the entire period till early May. We can easily pinpoint the inception of the campaign on Twitter, which occurs in the afternoon of April 30, 2017. After that, a surge in the volume of tweets, peaking at nearly 300 per minute, happens in the run up to

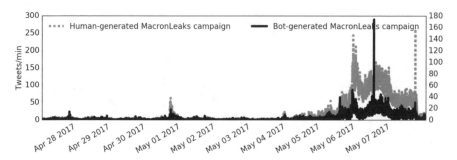

Fig. 5 Timeline of the volume of tweets generated every minute, respectively by human users (dashed grey line) and social bots (solid purple line), between April 27 and May 7, 2017, and related to MacronLeaks. Spikes in bot-generated content often slightly precedes spikes in human posts, suggesting that bots can trigger cascades of disinformation (from [28])

Election Day, between May 5 and May 6, 2017. It is worth noting that such a peak is nearly comparable in scale to the volume of the regular discussion, suggesting that for a brief interval of time (roughly 48 h) the MacronLeaks disinformation campaign acquired significant attention [28].

Summary: Bot Dynamics Like in the previous study, we here provide a characterization of the Twitter activity, this time specifically related to MacronLeaks, for both bot and human accounts. In Fig. 5, we show the timeline of the volume of tweets generated respectively by human users (dashed grey line) and bots (solid purple line), between April 27 and May 7, 2017, and related to MacronLeaks. The amount of activity is substantially close to zero until May 5, 2017, in line with the first coordination efforts as well as the information leaks spurred from other social platforms, as discussed in the paper [28]. Spikes in bot-generated content often appear to slightly precede spikes in human posts, suggesting that bots can trigger cascades of disinformation [67]. At peak, the volume of bot-generated tweets is comparable with the that of human-generated ones. Further investigation revealed that the users who engaged with bots pushed the *MacronLeaks* disinformation campaign were mostly foreigners with pre-existing interest in alt-right topics and alternative news media, rather than French users. Furthermore, we highlighted an anomalous account usage pattern where hundreds of bot accounts used in the 2017 French Presidential elections were also present in the 2016 US Presidential Election discussion, which suggested the possible existence of a black market for reusable political disinformation bots [28].

Summary: Sentiment Dynamics Identically to the 2016 US Presidential Election study, we annotated all tweets in this corpus using *SentiStrength*, and subsequently studied the evolution of the sentiment of tweets in the 2017 French Presidential Election discussion. Figure 6 shows the temporal distribution of tweets' sentiment disaggregated by intensity: the four panels illustrate the overall timeline of the volume of tweets that exhibit positive and negative sentiment at the hourly resolution, for sentiment polarities ranging from 1 (lowest) to 4 (highest) in both positive

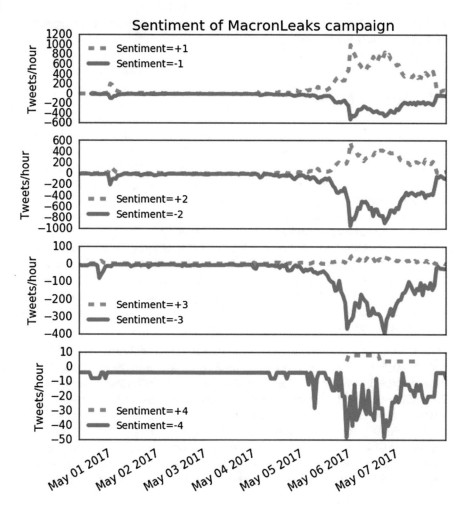

Fig. 6 Temporal distribution of sentiment disaggregated by sentiment intensity (hourly resolution). The sign on the y-axis captures the amount of tweets in the positive (resp., negative) sentiment dimension

and negative spectra. What appears evident is that, as Election Day approaches, moderately and highly negative tweets (sentiment scores of −2, −3, and −4) significantly outnumber the moderately and highly positive tweets, at times by almost an order of magnitude. For example, between May 6 and 7, 2017, on average between 300 and 400 tweets with significant negative sentiment (sentiment scores of −3) were posted every hour, compared with an average of between 10 and 50 tweets with an equivalently positive sentiment (score scores of +3). Since the discussion during that period was significantly driven by bots, and bots focused against Macron, our analysis suggested that bots were pushing negative campaigns against that candidate aimed at smearing his credibility and weakening his position in the eve of the May 7's election.

3.3   2018 US Midterms

The notorious investigation on Russian interference led by the US Senate Select Committee on Intelligence (SSCI) put social media service providers (SMSPs) at the center-stage of the public debate. According to reports, SMPSs started to devote more efforts to "sanitize" their platforms, including ramping up the technological solutions to detect and fight abuse. Much attention has been devoted to identifying and suspending *inauthentic activity*, a term that captures a variety of tools used to carry out manipulation, including bot and troll accounts.

Hence, it is natural to ask whether these countermeasures proved effective, or if otherwise the strategies and technologies bots typically used until 2017 evolved, and to what extent they successfully adapted to the changing social media defenses and thus escaped detection. We recently set to answer these questions: to this purpose, we monitored and investigated the online activity surrounding the 2018 US Midterm elections what were held on November 6, 2018.

Data Collection   We collected data for six weeks, from October 6, 2018 to November 19, 2018, i.e., one month prior and until two weeks after election day. Tweets were collected using the Twitter Streaming API and following these keywords: *2018midtermelections*, *2018midterms*, *elections*, *midterm*, and *midtermelections*. Post-processing and cleaning procedures are described in detail in our paper [53]: we retained only tweets in English, and manually removed tweets that were out of context, e.g., tweets related to other countries' elections (Cameroon, Congo, Biafra, Kenya, India, etc.) that were present in our initial corpus because they contained the same keywords we tracked. The final dataset contains 2.6M tweets, posted by nearly 1M users.

Bot Detection   Similarly to the 2016 US Presidential election study, since this study was a post-mortem (i.e., not in real time but after the events), we adopted Botometer to infer the bot scores of the users in our dataset. The only distinction worth mentioning is that we used the Botometer API version *v3* that brings new features and a non-linear re-calibration of the model: in line with the associated study's recommendations [86], we used a threshold of 0.3 (which corresponds to a 0.5 threshold from previous versions of Botometer) to separate bots from humans (note that the results remain substantially unchanged if a higher threshold was used). As a result, we obtained that 21.1% of the accounts were categorized as bots, which were responsible for 30.6% of the total tweets in our dataset. Manual validation procedures assessed the reasonable quality of these annotations. The resulting evidence suggests that bots were still present, and accounted for a significant amount of the tweets posted in the context of the political discourse revolving around the 2018 US Midterms.

Interestingly, about 40 thousand accounts were already inactive at the time of our analysis, and thus we were not able to infer their bot scores using the Twitter API. We manually verified that 99.4% of them were suspended by Twitter, corroborating the hypothesis that these were bots as well, and were suspended by Twitter in the time between the events and our post-mortem analysis, which was carried out in early 2019.

Political Leaning Inference Next, we set to determine if bots exhibited a clear political leaning, and if they acted according to that preference. To label accounts as conservative or liberal, we used a label propagation approach that leveraged the political alignment of news sources whose URLs were posted by the accounts in the dataset. Lists of partisan media outlets were taken from third-party organizations, namely AllSides.Org and MediaBiasFactCheck.Com. The details of our label propagation algorithm are explained in our paper [53]. Ultimately, the procedure allowed us to reliably infer, with accuracy above 89%, the political alignment of the majority of human and bot accounts in our corpus. These were factored into the subsequent analyses aimed at determining partisan strategies and narratives (see [53]).

Summary: Bot Activity and Strategies Provided the evidence that bots were still present despite the efforts of the SMSPs to sanitize their platforms, we aimed at determining the degree to which they were embedded in the human ecosystem, specifically in the retweet network. This network is of central importance in our analysis, because it conveys information diffusion dynamics; many recent studies suggested a connection between bots and the spread of unverified and false information [67, 83]. It is therefore of paramount importance to determine if bots still played a role in the retweet network of election-related social media discourse as of 2018.

To this aim, we resorted to perform the k-core decomposition analysis. In social network theory, a k-core is a subgraph of a graph where all nodes have degree at least equal to k. The intuition is that, as k grows, one is looking at increasingly more highly-connected nodes' subgraphs. Evidence suggests that high k-cores are associated with nodes that are more embedded, thus influential, for the network under investigation [84].

If bots were still influential in the 2018 US Midterm election discussion, our hypothesis is that we would find them in high concentration predominantly into high k cores. This would be consistent with our findings related to the 2016 US Presidential Election discussion [10].

Figure 7 corroborates our intuition. Specifically, we show the percentage of both conservative and liberal human and bot accounts as a function of varying k. Two patterns are worth discussing: first, as k increases, the fraction of conservative bots grows, while the prevalence of liberal bots remains more or less constant; conversely, the prevalence of human accounts decreases, with growing k, more markedly for liberal users than conservative ones. We summarize these findings suggesting that conservative bots were situated in a premium position in the retweet network, and therefore may have affected information spread [53].

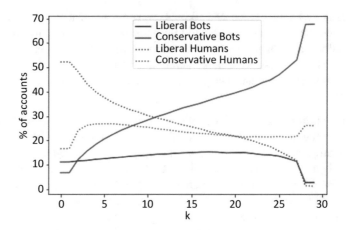

Fig. 7 K-core decomposition: liberal vs. conservative bots and humans (from [53])

3.4   2016 vs 2018: A Comparative Bot Analysis

Having identified and analyzed the activity of human and bot accounts in the context
of the political discourse associated to US election events in both 2016 and 2018,
it is natural to ask whether these studies involved a similar set of accounts. In other
words, it is worth determining whether there exists a continuum of users that are
active in both time periods under investigation. If this is the case, it would be
interesting to study the users present in both periods, determine whether any of
them are the bots under scrutiny in the previous studies, and ultimately understand
if the strategies they may have exhibited evolved, possibly to escape detection or
avoid further scrutiny of SMSPs.

Data Collection To answer the questions above, we isolated the users present
in both the 2016 and 2018 datasets described above. This process yielded over
278 thousand accounts, active in both periods. Further processing and cleaning
procedures, as detailed in our paper [54], brought the dataset down to 245 K users,
accounting for over 8.3 M tweets in 2016 and 660 K in 2018. Botometer was used
to determine the bot scores of these accounts. As a result, 12.6% of these accounts
scored high in bot scores and were therefore classified as bots. We used this dataset
to study the evolution of behavior of bots over the time period of study.

Summary: Bot Evolution Dynamics One advantage of bots over humans is their
scalability. Since bots are controlled by software rather than human users, as such
they can work over the clock, they don't need to take rests and don't have the finite
cognitive capacity and bandwidth that dictates how humans operate on social media
[63]. In principle, a bot could post continuously without any break, or at regular yet
tight intervals of time. As a matter of fact, primitive bots used these simple strategies
[58, 65]. However, such obvious patterns are easy to spot automatically, hence not

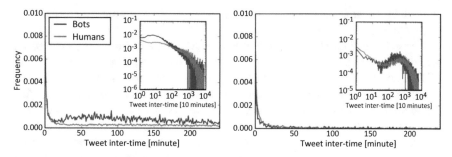

Fig. 8 Tweet inter-event time by bots and humans in 2016 (left) and 2018 (right). A clear distinction in temporal signature between bots and humans was evident in 2016, but vanished in 2018 (from [54])

very effective. There is therefore a trade-off between realistic-looking activity and effectiveness. In other words, one can investigate the patterns of *inter-event time* betweet a tweet post and its subsequent, and lay out the frequency distribution in an attempt to distill the difference between human and bot accounts' temporal dynamics.

Figure 8 illustrates the tweet inter-time distribution by bots and humans in 2016 (left) and 2018 (right). It is apparent that, while in 2016 bots exhibited a significantly different frequency distribution with respect to their human counterparts, in 2018 this distinction has vanished. In fact, statistical testing of distribution differences suggests that human and bot temporal signatures are indistinguishable in 2018. The discrepancy is particularly relevant in the time range between 10 min and 3 h, consistent with other findings [63]: in 2016, bots shared content at a higher rate with respect to human users.

Our work [54] corroborates the hypothesis that bots are continuously changing and evolving to escape detection. Further examples that we reported also illustrate other patterns of behavior that have changed between 2016 and 2018: for instance, the sentiment that was expressed in favor or against political candidates in 2018 reflects significantly better what the human crowd is expressing. However, in 2016, bots' sentiment drastically diverged, in a manner easy to detect, from that of the human's counterpart, as we discussed earlier.

4 Conclusions

In this chapter, we set to discuss our latest results regarding the role of bots within online political discourse in association with three major political events.

First, we described the results of our analysis that unveiled a significant amount of bots distorting the online discussion in relation to the 2016 US Presidential election. We characterized the activities of such bots, and illustrated how they successfully fostered interactions by means of retweets at the same rate human users did. Other researchers suggested that this played a role in the spread of false news during that time frame [67, 83].

Second, we highlighted the role of bots in pushing a disinformation campaign, known as *MacronLeaks*, in the run up to the 2017 French Presidential election. We demonstrated how it is possible to easily pinpoint the inception of this disinformation campaign on Twitter, and we illustrated how its popularity peak was comparable with that of regular political discussion. We also hypothesized that this disinformation campaign did not have a major success in part because it was tailored around the information needs and usage patterns of the American alt-right community rather than French-speaking audience. Moreover, we found that several hundreds of bot accounts were re-purposed from the 2016 US Election. Ultimately, we suggested the possibility that a black market for reusable political bots may exist [28].

Third, we studied the 2018 US Midterms, to investigate if bots were still present and active. Our analysis illustrated that not only bots were almost as prevalent as in the two other events, but also that conservative bots played a central role in the highly-connected core of the retweet network. These findings further motivated a comparative analysis contrasting the activity of bots and humans in 2016 and 2018. Our study highlighted that a core of over 245 K users, of which 12.1% were bots, was active in both events. Our results suggest that bots may have evolved to better mimic human temporal patterns of activity.

With the increasing sophistication of Artificial Intelligence, the ability of bots to mimic human behavior to escape detection is greatly enhanced. This poses challenges for the research community, specifically in the space of bot detection. Whether it is possible to win this arms race is yet to be determined: any party with significant resources can deploy state of the art technologies to enact influence operations and other forms of manipulation of public opinion.

The availability of powerful neural language models lowers the bar to adopt techniques that allow to build credible bots. For example, it may be already in principle possible to automatize almost completely the generation of genuine-looking text. This may be used to push particular narratives, to artificially build traction for political arguments that may otherwise have little or no human organic support.

Ultimately, the evidence that our studies, and the work of many other researchers in this field, have brought strongly suggest that more policy and regulations may be warranted, and that technological solutions alone may not be sufficient to tackle the issues of bot interference in political discourse.

Acknowledgements The author is grateful to his collaborators and coauthors on the topics covered in this paper, in particular Adam Badawy, Alessandro Bessi, Ashok Deb, and Luca Luceri, who contributed significantly to three papers widely discussed in this chapter [10, 53, 54].

References

1. Abokhodair, N., Yoo, D., McDonald, D.W.: Dissecting a social botnet: growth, content and influence in twitter. In: Proceedings of the 18th ACM Conference on Computer Supported Cooperative Work & Social Computing, pp. 839–851. ACM (2015)
2. Adamic, L.A., Glance, N.: The political blogosphere and the 2004 us election: divided they blog. In: 3rd International Workshop on Link Discovery, pp. 36–43. ACM (2005)
3. Alarifi, A., Alsaleh, M., Al-Salman, A.: Twitter turing test: identifying social machines. Inf. Sci. **372**, 332–346 (2016)
4. Allem, J.-P., Ferrara, E., Uppu, S.P., Cruz, T.B., Unger, J.B.: E-cigarette surveillance with social media data: social bots, emerging topics, and trends. JMIR Public Health Surveill. **3**(4), e98 (2017)
5. Aral, S., Walker, D.: Creating social contagion through viral product design: a randomized trial of peer influence in networks. Manag. Sci. **57**(9), 1623–1639 (2011)
6. Badawy, A., Ferrara, E., Lerman, K.: Analyzing the digital traces of political manipulation: the 2016 russian interference twitter campaign. In: Proceedings of the 2018 IEEE/ACM International Conference on Advances in Social Networks Analysis and Mining 2018 (2018)
7. Badawy, A., Lerman, K., Ferrara, E.: Who falls for online political manipulation? In: Companion Proceedings of the 2019 World Wide Web Conference, pp. 162–168 (2019)
8. Barberá, P., Wang, N., Bonneau, R., Jost, J.T., Nagler, J., Tucker, J., González-Bailón, S.: The critical periphery in the growth of social protests. PLoS One **10**(11), e0143611 (2015)
9. Bekafigo, M.A., McBride, A.: Who tweets about politics? political participation of twitter users during the 2011 gubernatorial elections. Soc. Sci. Comput. Rev. **31**(5), 625–643 (2013)
10. Bessi, A., Ferrara, E.: Social bots distort the 2016 us presidential election online discussion. First Monday **21**(11) (2016)
11. Boshmaf, Y., Muslukhov, I., Beznosov, K., Ripeanu, M.: The socialbot network: when bots socialize for fame and money. In: Proceedings of the 27th Annual Computer Security Applications Conference, pp. 93–102. ACM (2011)
12. Boyd, D., Crawford, K.: Critical questions for big data: provocations for a cultural, technological, and scholarly phenomenon. Inf. Commun. Soc. **15**(5), 662–679 (2012)
13. Carlisle, J.E., Patton, R.C.: Is social media changing how we understand political engagement? an analysis of facebook and the 2008 presidential election. Polit. Res. Q. **66**(4), 883–895 (2013)
14. Centola, D.: An experimental study of homophily in the adoption of health behavior. Science **334**(6060), 1269–1272 (2011)
15. Cha, M., Haddadi, H., Benevenuto, F., Gummadi, K.P.: Measuring user influence in twitter: the million follower fallacy. In: Fourth International AAAI Conference on Weblogs and Social Media (ICWSM 2010), pp. 10–17. AAAI Press (2010)
16. Chavoshi, N., Hamooni, H., Mueen, A.: Debot: twitter bot detection via warped correlation. In: ICDM, pp. 817–822 (2016)
17. Conover, M., Ratkiewicz, J., Francisco, M.R., Gonçalves, B., Menczer, F., Flammini, A.: Political polarization on twitter. ICWSM **133**, 89–96 (2011)
18. Conover, M.D., Davis, C., Ferrara, E., McKelvey, K., Menczer, F., Flammini, A.: The geospatial characteristics of a social movement communication network. PLoS One **8**(3), e55957 (2013)
19. Conover, M.D., Ferrara, E., Menczer, F., Flammini, A.: The digital evolution of occupy wall street. PLoS One **8**(5), e64679 (2013)
20. Cresci, S., Di Pietro, R., Petrocchi, M., Spognardi, A., Tesconi, M.: The paradigm-shift of social spambots: evidence, theories, and tools for the arms race. In: Proceedings of the 26th International Conference on World Wide Web Companion, pp. 963–972. International World Wide Web Conferences Steering Committee (2017)
21. Davis, C.A., Varol, O., Ferrara, E., Flammini, A., Menczer, F.: Botornot: a system to evaluate social bots. In: WWW '16 Companion Proceedings of the 25th International Conference Companion on World Wide Web, pp. 273–274. ACM (2016)

22. De Cristofaro, E., Kourtellis, N., Leontiadis, I., Stringhini, G., Zhou, S., et al.: Lobo: evaluation of generalization deficiencies in twitter bot classifiers. In: Proceedings of the 34th Annual Computer Security Applications Conference, pp. 137–146. ACM (2018)
23. DiGrazia, J., McKelvey, K., Bollen, J., Rojas, F.: More tweets, more votes: social media as a quantitative indicator of political behavior. PLoS One 8(11), e79449 (2013)
24. Echeverria, J., Besel, C., Zhou, S.: Discovery of the twitter Bursty botnet. Data Science for Cyber-Security (2017)
25. Effing, R., Van Hillegersberg, J., Huibers, T.: Social media and political participation: are facebook, twitter and youtube democratizing our political systems? In International Conference on Electronic Participation, pp. 25–35. Springer (2011)
26. El-Khalili, S.: Social media as a government propaganda tool in post-revolutionary Egypt. First Monday 18(3) (2013)
27. Ferrara, E.: Manipulation and abuse on social media. ACM SIGWEB Newsletter (4) (2015)
28. Ferrara, E.: Disinformation and social bot operations in the run up to the 2017 french presidential election. First Monday 22(8) (2017)
29. Ferrara, E.: The history of digital spam. Commun. ACM 62(8), 2–91 (2019)
30. Ferrara, E., Varol, O., Davis, C., Menczer, F., Flammini, A.: The rise of social bots. Commun. ACM 59(7), 96–104 (2016)
31. Ferrara, E., Varol, O., Menczer, F., Flammini, A.: Detection of promoted social media campaigns. In: Tenth International AAAI Conference on Web and Social Media, pp. 563–566 (2016)
32. Ferrara, E., Yang, Z.: Measuring emotional contagion in social media. PLoS One 10(11), e0142390 (2015)
33. Ferrara, E., Yang, Z.: Quantifying the effect of sentiment on information diffusion in social media. PeerJ Comput. Sci. 1, e26 (2015)
34. Gao, H., Barbier, G., Goolsby, R.: Harnessing the crowdsourcing power of social media for disaster relief. IEEE Intell. Syst. 26(3), 10–14 (2011)
35. Gao, H., Hu, J., Wilson, C., Li, Z., Chen, Y., Zhao, B.Y.: Detecting and characterizing social spam campaigns. In: Proceedings of the 10th ACM SIGCOMM Conference on Internet Measurement, pp. 35–47. ACM (2010)
36. Gilani, Z., Kochmar, E., Crowcroft, J.: Classification of twitter accounts into automated agents and human users. In: Proceedings of the 2017 IEEE/ACM International Conference on Advances in Social Networks Analysis and Mining 2017, pp. 489–496. ACM (2017)
37. González-Bailón, S., Borge-Holthoefer, J., Moreno, Y.: Broadcasters and hidden influentials in online protest diffusion. Am. Behav. Sci. 57(7), 943–965 (2013)
38. González-Bailón, S., Borge-Holthoefer, J., Rivero, A., Moreno, Y.: The dynamics of protest recruitment through an online network. Sci. Rep. 1, 197 (2011)
39. Howard, P.N.: New Media Campaigns and the Managed Citizen. Cambridge University Press, Cambridge (2006)
40. Howard, P.N., Kollanyi, B.: Bots, #strongerin, and #brexit: computational propaganda during the uk-eu referendum. Available at SSRN 2798311 (2016)
41. Hwang, T., Pearce, I., Nanis, M.: Socialbots: voices from the fronts. Interactions 19(2), 38–45 (2012)
42. Jagatic, T.N., Johnson, N.A., Jakobsson, M., Menczer, F.: Social phishing. Commun. ACM 50(10), 94–100 (2007)
43. Jin, X., Lin, C., Luo, J., Han, J.: A data mining-based spam detection system for social media networks. Proceedings VLDB Endowment 4(12), 1458–1461 (2011)
44. Kollanyi, B., Howard, P.N., Woolley, S.C.: Bots and automation over twitter during the first us presidential debate. Technical report, COMPROP Data Memo (2016)
45. Kramer, A.D., Guillory, J.E., Hancock, J.T.: Experimental evidence of massive-scale emotional contagion through social networks. Proc. Natl. Acad. Sci. 111(24), 8788–8790 (2014)
46. Kudugunta, S., Ferrara, E.: Deep neural networks for bot detection. Inf. Sci. 467, 312–322 (2018)

47. Kümpel, A.S., Karnowski, V., Keyling, T.: News sharing in social media: a review of current research on news sharing users, content, and networks. Soc. Media Soc. **1**(2), 2056305115610141 (2015)
48. Kwak, H., Lee, C., Park, H., Moon, S.: What is twitter, a social network or a news media? In: Proceedings of the 19th International Conference on World Wide Web, pp. 591–600 (2010)
49. Latonero, M., Shklovski, I.: Emergency management, twitter, and social media evangelism. In: Using Social and Information Technologies for Disaster and Crisis Management, pp. 196–212. IGI Global, Hershey (2013)
50. Lazer, D., Pentland, A.S., Adamic, L., Aral, S., Barabasi, A.L., Brewer, D., Christakis, N., Contractor, N., Fowler, J., Gutmann, M., et al.: Life in the network: the coming age of computational social science. Science (New York, NY) **323**(5915), 721 (2009)
51. Lee, K., Caverlee, J., Webb, S.: The social honeypot project: protecting online communities from spammers. In: Proceedings of the 19th International Conference on World Wide Web, pp. 1139–1140. ACM (2010)
52. Lee, K., Caverlee, J., Webb, S.: Uncovering social spammers: social honeypots+ machine learning. In: Proceedings of the 33rd International ACM SIGIR Conference on Research and Development in Information Retrieval, pp. 435–442. ACM (2010)
53. Luceri, L., Deb, A., Badawy, A., Ferrara, E.: Red bots do it better: comparative analysis of social bot partisan behavior. In: Companion Proceedings of the 2019 World Wide Web Conference, pp. 1007–1012 (2019)
54. Luceri, L., Deb, A., Giordano, S., Ferrara, E.: Evolution of bot and human behavior during elections. First Monday **24**(9) (2019)
55. Lutz, C., Hoffmann, C.P., Meckel, M.: Beyond just politics: a systematic literature review of online participation. First Monday **19**(7) (2014)
56. Markines, B., Cattuto, C., Menczer, F.: Social spam detection. In: Proceedings of the 5th International Workshop on Adversarial Information Retrieval on the Web, pp. 41–48 (2009)
57. Messias, J., Schmidt, L., Oliveira, R., Benevenuto, F.: You followed my bot! transforming robots into influential users in twitter. First Monday **18**(7) (2013)
58. Metaxas, P.T., Mustafaraj, E.: Social media and the elections. Science **338**, 472–473 (2012)
59. Minnich, A., Chavoshi, N., Koutra, D., Mueen, A.: Botwalk: efficient adaptive exploration of twitter bot networks. In: Proceedings of the 2017 IEEE/ACM International Conference on Advances in Social Networks Analysis and Mining 2017, pp. 467–474. ACM (2017)
60. Mønsted, B., Sapieżyński, P., Ferrara, E., Lehmann, S.: Evidence of complex contagion of information in social media: an experiment using twitter bots. PLos One **12**(9), e0184148 (2017)
61. Mukherjee, A., Liu, B., Glance, N.: Spotting fake reviewer groups in consumer reviews. In: Proceedings of the 21st International Conference on World Wide Web, pp. 191–200 (2012)
62. Mustafaraj, E., Metaxas, P.T.: From obscurity to prominence in minutes: political speech and real-time search (2010)
63. Pozzana, I., Ferrara, E.: Measuring bot and human behavioral dynamics. arXiv preprint arXiv:1802.04286 (2018)
64. Radford, A., Wu, J., Child, R., Luan, D., Amodei, D., Sutskever, I.: Language models are unsupervised multitask learners. OpenAI Blog **1**(8) (2019)
65. Ratkiewicz, J., Conover, M., Meiss, M., Gonçalves, B., Flammini, A., Menczer, F.: Detecting and tracking political abuse in social media. ICWSM **11**, 297–304 (2011)
66. Ratkiewicz, J., Conover, M., Meiss, M., Gonçalves, B., Patil, S., Flammini, A., Menczer, F.: Truthy: mapping the spread of astroturf in microblog streams. In: Proceedings of the 20th International Conference Companion on World Wide Web, pp. 249–252. ACM (2011)
67. Shao, C., Ciampaglia, G.L., Varol, O., Yang, K.-C., Flammini, A., Menczer, F.: The spread of low-credibility content by social bots. Nat. Commun. **9**(1), 4787 (2018)
68. Shorey, S., Howard, P.N.: Automation, algorithms, and politics| automation, big data and politics: a research review. Int. J. Commun. **10**, 24 (2016)
69. Song, J., Lee, S., Kim, J.: Spam filtering in twitter using sender-receiver relationship. In: International Workshop on Recent Advances in Intrusion Detection, pp. 301–317 (2011)

70. Stella, M., Ferrara, E., De Domenico, M.: Bots increase exposure to negative and inflammatory content in online social systems. Proc. Natl. Acad. Sci. **115**(49), 12435–12440 (2018)
71. Stieglitz, S., Brachten, F., Ross, B., Jung, A.-K.: Do social bots dream of electric sheep? a categorisation of social media bot accounts. arXiv preprint arXiv:1710.04044 (2017)
72. Stringhini, G., Kruegel, C., Vigna, G.: Detecting spammers on social networks. In: Proceedings of the 26th Annual Computer Security Applications Conference, pp. 1–9. ACM (2010)
73. Stukal, D., Sanovich, S., Bonneau, R., Tucker, J.A.: Detecting bots on russian political twitter. Big Data **5**(4), 310–324 (2017)
74. Subrahmanian, V., Azaria, A., Durst, S., Kagan, V., Galstyan, A., Lerman, K., Zhu, L., Ferrara, E., Flammini, A., Menczer, F.: The darpa twitter bot challenge. Computer **49**(6), 38–46 (2016)
75. Sutton, J.N., Palen, L., Shklovski, I.: Backchannels on the Front Lines: Emergency Uses of Social Media in the 2007 Southern California Wildfires. University of Colorado, Boulder (2008)
76. Thelwall, M., Buckley, K., Paltoglou, G., Cai, D., Kappas, A.: Sentiment strength detection in short informal text. J. Am. Soc. Inf. Sci. Tech. **61**(12), 2544–2558 (2010)
77. Theocharis, Y., Lowe, W., van Deth, J.W., García-Albacete, G.: Using twitter to mobilize protest action: online mobilization patterns and action repertoires in the occupy wall street, indignados, and aganaktismenoi movements. Inf. Commun. Soc. **18**(2), 202–220 (2015)
78. Thomas, K., Grier, C., Song, D., Paxson, V.: Suspended accounts in retrospect: an analysis of twitter spam. In: Proceedings of the 2011 ACM SIGCOMM Conference on Internet Measurement Conference, pp. 243–258. ACM (2011)
79. Thomas, K., McCoy, D., Grier, C., Kolcz, A., Paxson, V.: Trafficking fraudulent accounts: the role of the underground market in twitter spam and abuse. In: Usenix Security, vol. 13, pp. 195–210 (2013)
80. Varol, O., Ferrara, E., Davis, C., Menczer, F., Flammini, A.: Online human-bot interactions: detection, estimation, and characterization. In: International AAAI Conference on Web and Social Media (2017)
81. Varol, O., Ferrara, E., Menczer, F., Flammini, A.: Early detection of promoted campaigns on social media. EPJ Data Sci. **6**(1), 13 (2017)
82. Varol, O., Ferrara, E., Ogan, C.L., Menczer, F., Flammini, A.: Evolution of online user behavior during a social upheaval. In: Proceedings 2014 ACM Conference on Web Science, pp. 81–90 (2014)
83. Vosoughi, S., Roy, D., Aral, S.: The spread of true and false news online. Science **359**(6380), 1146–1151 (2018)
84. Wasserman, S., Faust, K.: Social Network Analysis: Methods and Applications, vol. 8. Cambridge university press, New York (1994)
85. Yang, C., Harkreader, R., Zhang, J., Shin, S., Gu, G.: Analyzing spammers' social networks for fun and profit: a case study of cyber criminal ecosystem on twitter. In: Proceedings of the 21st International Conference on World Wide Web, pp. 71–80. ACM (2012)
86. Yang, K.-C., Varol, O., Davis, C.A., Ferrara, E., Flammini, A., Menczer, F.: Arming the public with artificial intelligence to counter social bots. Hum. Behav. Emerg. Technol. **1**, e115 (2019)
87. Yang, X., Chen, B.-C., Maity, M., Ferrara, E.: Social politics: agenda setting and political communication on social media. In: International Conference on Social Informatics, pp. 330–344. Springer (2016)
88. Yates, D., Paquette, S.: Emergency knowledge management and social media technologies: a case study of the 2010 haitian earthquake. Int. J. Inf. Manag. **31**(1), 6–13 (2011)
89. Yin, J., Lampert, A., Cameron, M., Robinson, B., Power, R.: Using social media to enhance emergency situation awareness. IEEE Intell. Syst. **27**(6), 52–59 (2012)
90. Zhang, X., Zhu, S., Liang, W.: Detecting spam and promoting campaigns in the twitter social network. In: Data Mining (ICDM), 2012 IEEE 12th International Conference on, pp. 1194–1199. IEEE (2012)

Part II
Techniques on Detecting and Mitigating Disinformation

Tensor Embeddings for Content-Based Misinformation Detection with Limited Supervision

Sara Abdali, Gisel G. Bastidas, Neil Shah, and Evangelos E. Papalexakis

Abstract Web-based technologies like social media have become primary news outlets for many people in recent years. Considering the fact that these digital outlets are extremely vulnerable to misinformation and fake news which may impact a user's opinion toward social, political, and economic issues, the necessity of robust and efficient approaches for misinformation detection task comes to light more than ever. The majority of misinformation detection approaches previously proposed leverage manually extracted features and supervised classifiers which require a large number of labeled data which is often infeasible to collect in practice. To meet this challenge, in this work we propose a novel strategy mixing tensor-based modeling of article content and semi-supervised learning on article embeddings for the misinformation detection task which requires very few labels to achieve state-of-the-art results. We propose and experiment with three different article content modeling variations which target article body text or title, and enable meaningful representations of word co-occurrences which are discriminative in the downstream news categorization task. We tested our approach on real world data and the evaluation results show that we achieve 75% accuracy using only 30% of the labeled data of a public dataset while the previously proposed and published SVM-based classifier results in 67% accuracy. Moreover, our approach achieves 71% accuracy on a large dataset using only 2% of the labels. Additionally, our approach is able to classify articles into different fake news categories (clickbait, bias, rumor, hate, and

The authors contributed equally to this paper.

S. Abdali (✉) · G. G. Bastidas · E. E. Papalexakis
University of California, Riverside, CA, USA
e-mail: sabda005@ucr.edu; gbast001@ucr.edu; epapalex@cs.ucr.edu

N. Shah
Snap Inc., Los Angeles, CA, USA
e-mail: nshah@snap.com

junk science) by only using the titles of the articles, with roughly 70% accuracy and 30% of the labeled data.

Keywords Misinformation · Fake news detection · Tensor decomposition · Semi-supervised learning

Misinformation propagation on the web, and especially via social media, is one of the most challenging problems due to the negative impact it may have on a user's opinions and their decision making process in political, economic, or social contexts. The spread of misinformation on Twitter during Hurricane Sandy in 2012 [8], the Boston Marathon blasts in 2013 [7] and US Presidential Elections on Facebook in 2016 [32] are some real world examples of misinformation propagation and its consequences which confirm the necessity of misinformation detection task.

As this is a very challenging and important problem, researchers have proposed different approaches and leveraged different signals to differentiate between misinformation and true facts. Among variety of signals researchers proposed to use so far, the content of under investigation case seems to convey more important information than other signals such as source and metadata information. So, devoting a lot of time to develop content based approaches not only makes a lot of sense in terms of detection of false parts but also enables us to design methods for fixing those parts as well.

Content based approaches for finding misinformation and fake news detection can be divided into two main categories. First, category approaches that focus on the text of the article. Second, category approaches that mainly focus on the title of the article specially when the task is to detect clickbaits [27, 28].

So far, many different strategies have been implemented to extract insightful information from content of the articles. For instance, there are many works on extracting linguistic based information like lexical features which comprise character and word level information and syntactic features that leverage sentence level features. There are some other approaches which exploit other content based information rather than linguistic based features e.g. number of nouns, proportion of positive/negative words article length and so on. The main drawback of lexical based approaches like bag of words is that there is no consideration about the relationship between the words within the text and the latent patterns they may form when they co-occur.

Moreover, the majority of works have been done so far use a supervised classifier for discriminate fake news and misinformation. For instance, Rubin et al.[24] used a linguistic features and a SVM-based classifier. In a similar way, Horne et al. [13] exploited a SVM classifier which was fed with stylistic and psychological features. The main issue with this category of misinformation detection approaches is that they mostly require a considerable amount of labeled data known as ground truth for training and testing. In reality, these labels are very limited and insufficient. Although there are a couple of fact checking websites like PolitiFact, FactCheck,

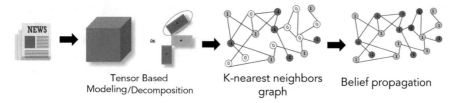

Fig. 1 Our proposed method discerns real from misinformative news articles via leveraging tensor representation and semi-supervised learning in graphs

Snopes and so on and so forth, they all require human expert and the task of fact checking is often a costly and time consuming process.

In contrast to aforementioned works, in this work, we propose a novel modeling approach which not only considers the relationship between words within article text but also provides us with latent patterns these relationship form for news article categories. Moreover, our method needs a limited amount of labeled data and human supervision. Figure 1 demonstrates a schematic representation of our proposed method.

Our main contributions are:

- We propose three different tensor-based embeddings to model content-based information of news articles which decomposition of these tensor-based models produce concise representations of spatial context and provide us with insightful patterns we leverage for classification of news articles.
- Introducing a tensor-based modeling approach which is not only applicable on body text but also capable of modeling news articles just using headlines, specially in situations in which the body text of the articles no longer exist.
- We leverage a propagation based approach for semi-supervised classification of news articles which enables us to classify news articles when there is scarcity of labels.
- We create a large dataset of misinformation and real news articles out of publicly shared tweets on Twitter.
- We evaluate our method on real datasets. Experiments on two previously used datasets demonstrate that our method outperforms prior works since it requires a fewer number of known labels and achieves comparable performance.

This book chapter is an extension of our previously published work [6] and the contributions of this extension are the study of the effect of using a higher dimension tensor-based model on binary classification task, strength of proposed method for multi-class classification of misinformation into different categories of fake news using the headlines of the articles, and the study of the sensitivity of the tensor based model to the length and categorical distribution of fake articles.

The remainder of this paper is organized as follows:

In Sect. 1, we describe preliminaries and definitions used throughout this book chapter. Section 2 defines the problem we aim to address. In Sect. 3, we introduce our proposed method in detail. In Sect. 4, we describe the implementation and the datasets we experimented on, then in Sect. 5 the experimental results and performance of our approach are described in detail. Section 6 involves sensitivity analysis, where we discuss the impact of categorical distribution and the length of news articles on the performance of our proposed method. Section 7 discusses about the related works, and finally, in Sect. 8 we conclude.

1 Preliminaries and Notations

In this section, we provide preliminary definitions for technical concepts that we use throughout this paper.

1.1 CP/PARAFAC Tensor Decomposition

A tensor is a multi-way array where its dimensions are referred to as modes. In other words, a tensor is an array with three or more than three indices. In linear algebra there is a factorization algorithm known as Singular Value Decomposition (SVD) in which we can factorize matrix X into the product of three matrices as follows:

$$X \approx \mathbf{U}\boldsymbol{\Sigma}\mathbf{V}^{T}$$

where the columns of U and V are orthonormal and the matrix $\boldsymbol{\Sigma}$ is a diagonal with positive real entries.

In fact, using SVD we can represent a matrix as a summation of rank 1 matrices:

$$X \approx \Sigma_{r=1}^{R}\sigma_r\mathbf{u}_r \circ \mathbf{v}_r$$

The Canonical Polyadic (CP) or PARAFAC decomposition is the simplest extension of SVD for higher mode matrices i.e, tensors [12] (Fig. 2). Indeed, CP/PARAFAC factorizes a tensor into a sum of rank-one tensors. For instance, a three-mode tensor is decomposed into a sum of outer products of three vectors as follows:

$$\mathcal{X} \approx \Sigma_{r=1}^{R}\mathbf{a}_r \circ \mathbf{b}_r \circ \mathbf{c}_r$$

where $\mathbf{a}_r \in \mathbb{R}^{I}$, $\mathbf{b}_r \in \mathbb{R}^{J}$, $\mathbf{c}_r \in \mathbb{R}^{K}$ and the outer product is given by [21, 31]:

$$(\mathbf{a}_r, \mathbf{b}_r, \mathbf{c}_r)(i, j, k) = \mathbf{a}_r(i)\mathbf{b}_r(j)\mathbf{c}_r(k)\forall i, j, k$$

Fig. 2 CP/PARAFAC decomposition of a 3-mode tensor

The factor matrices are defined as $\mathbf{A} = [\mathbf{a}_1\ \mathbf{a}_2 \dots \mathbf{a}_R]$, $\mathbf{B} = [\mathbf{b}_1\ \mathbf{b}_2 \dots \mathbf{b}_R]$, and $\mathbf{C} = [\mathbf{c}_1\ \mathbf{c}_2 \dots \mathbf{c}_R]$ where $\mathbf{A} \in \mathbb{R}^{I \times R}$, $\mathbf{B} \in \mathbb{R}^{J \times R}$, and $\mathbf{C} \in \mathbb{R}^{K \times R}$ denote the factor matrices and R is the rank of the decomposition or the number of columns in the factor matrices. So, the optimization problem for finding factor matrices is as follows:

$$\min_{A,B,C} = \| \mathcal{X} - \Sigma_{r=1}^{R} \mathbf{a}_r \circ \mathbf{b}_r \circ \mathbf{c}_r \|^2$$

One of the simplest and most effective approaches for solving optimization problem above is to use Alternating Least Squares (ALS) which solves for any of the factor matrices by fixing the others [21, 31].

1.2   k-Nearest-Neighbor Graph

A k-nearest-neighbor, or k-NN graph is a graph in which node p and node q are connected by an edge if node p is in q's k-nearest-neighborhood or node q is in p's k-nearest-neighborhood. k is a hyperparameter that determines the density of the graph – thus, a graph constructed with small k may be sparse or poorly connected.

The k-nearest-neighbors of a point in n-dimensional space are defined using a "closeness" relation where proximity is often defined in terms of a distance metric [10] such as Euclidean ℓ_2 distance. Thus, given a set of points $\mathcal{P}$ in n-dimensional space, a k-NN graph on $\mathcal{P}$ can be constructed by computing the ℓ_2 distance between each pair of points and connecting each point with the k most proximal ones.

The ℓ_2 distance d between two points p and q in n-dimensional space is defined as:

$$d(p, q) = \sqrt{\sum_{i=1}^{n} (q_i - p_i)^2}$$

1.3   Belief Propagation

Belief propagation is a message passing algorithm usually applied for calculation of marginal distribution on graph based models such as Markov or Bayesian networks. Several different versions for belief propagation have been introduced each of which for a different graphical model. However, the iterative message passing mechanism throughout the network, is a common function used for all different versions of belief propagation because the operative intuition behind this algorithm is that the nodes which are "close" are more likely to have similar values known as "belief".

Suppose $m_{j \hookrightarrow i}(x_i)$ denote the message passes from node i to node j. $m_{j \hookrightarrow i}(x_i)$ conveys the opinion of node i about the belief of node j. Each node of a given graph G uses the messages received from neighboring nodes to compute its belief iteratively as follows:

$$b_i(x_i) \propto \prod_{j \in (N_i)} m_{j \hookrightarrow i}(x_i)$$

Where N_i denotes all the neighboring nodes of node i [3, 35].

In this work, we define the belief as the label of a news article and given a set of known labels, we use FaBP as a means to propagate label likelihood over the nearest neighbor graph.

Fast Belief Propagation (FaBP) [17] is a fast and linearized guilt-by-association method, which improves the basic idea of belief propagation (BP) we discussed.

The FaBP algorithm solves the following linear system:

$$[\mathbf{I} + a\mathbf{D} - c'\mathbf{A}]b_h = \phi_h$$

where ϕ_h and b_h denote prior and final beliefs, respectively. $\mathbf{A}$ denotes the $n \times n$ adjacency matrix of an underlying graph of n nodes, $\mathbf{I}$ denotes the $n \times n$ identity matrix, and $\mathbf{D}$ is a $n \times n$ diagonal matrix of degrees where $\mathbf{D}_{ii} = \sum_j \mathbf{A}_{ij}$ and $\mathbf{D}_{ij} = 0$ for $i \neq j$. Finally, we define $a = \frac{4h_h^2}{1 - 4h_h^2}$ and $c' = \frac{2h_h}{(1 - 4h_h^2)}$ where h_h denotes the homophily factor between nodes (i.e. their "coupling strength" or association). More specifically, higher homophily means that close nodes tend to have more similar labels. The coefficient values are set as above for convergence reasons; we refer the interested reader to [17] for further discussion.

2   Problem Definition

In this work, we follow the definition used in [27] and consider articles which are *"intentionally and verifiably false,"* as fake news or misinformation. Based on this definition, we aim to leverage content of news articles to discern fake news from real news. Henceforth, by "content," we refer to the body text or headline text of the articles.

Suppose we have $\mathcal{N} = \{n_1, n_2, n_3, \ldots, n_M\}$ a collection of M news articles each of which is a set of words. We create $\mathcal{D} = \{w_1, w_2, w_3, \ldots, w_I\}$ a dictionary of size I out of unique words of all articles. Assume the labels of some news articles are available. with aforementioned notations in mind, we define two problems as follows:

Problem 1 Given a collection of $\mathcal{N}$ news articles and a label vector l with entries labeled real, fake or unknown articles, the goal is to **predict** the labels of the unknown labels.

We address this problem as a binary classification problem in which a news article is classified either fake or real.

Problem 2 Given a collection of $\mathcal{N}$ news articles and a label vector l with entries labeled as real, bias, clickbait, conspiracy, fake, hate, junck science, satire, and unreliable or unknown articles, the goal is to **predict** the labels of the unknown labels.

We address this problem as a multi class classification problem in which a news article is classified as one of the categories mentioned above.

3   Proposed Method

In this section, we introduce a content-based method for semi-supervised classification of news articles. This method consists of three consecutive steps: step 1 refers to the modeling of articles' content as a tensor, and decomposition of resulted model into factor matrices. In step 2, we leverage the factor matrix corresponding to article mode to create a k-NN graph to represent the proximity of articles. In step 3, we use FaBP to propagate very few amount of labels throughout the graph in order to find the unknown labels. Figure 1 illustrates our proposed method. In what follows, we discuss each step in detail.

Step 1: Tensor Decomposition

The very first step of our proposed method is to model news articles. We propose a novel approach i.e., a tensor-based approach to model an article's content. To this end, we define 4 different models as follows:

Model 1: Term-Term-Article (TTA) Suppose there is a collection of M news articles $\mathcal{N} = \{n_1, n_2, n_3, \ldots, n_M\}$. We define each news article in $\mathcal{N}$ as a matrix representing the co-occurrence of the words within a sliding window of size w. We create a tensor embedding by stacking co-occurrence matrices of all news article in $\mathcal{N}$ as proposed in [14].

In other words, in this three-mode tensor $\mathcal{X} \in \mathbb{R}^{I \times I \times M}$ $(term, term, article)$ each news article is a co-occurrence matrix in which entry $(term_1, term_2)$ of the matrix is a binary value representing the co-occurrence of $term_1$ and $term_2$ (for binary-based model) or the number of times this pair of words co-occur throughout the text (for frequency-based model) within a window of size w (usually 5–10).[1]

Model 2: Term-Term-Term-Article (3TA) In previous model, we created a tensor embedding as a representative of all couples of co-occurred words within each news article. In this model, we aim at designing a tensor based embedding which enables us to demonstrate meaningful co-occurrence of larger set of words within an article, i.e., instead of pairwise co-occurrence we want to capture all triple-way co-occurred words Fig. 3. Therefore, the resulted tensor embedding is going to be a 4-mode tensor $\mathcal{X} \in \mathbb{R}^{I \times I \times I \times M}$ $(term, term, term, article)$ created by stacking 3-mode tensors each of which represents triple-way co-occurrence within each news article.

Model 3: Term-Term-Article on news Title (TTA out of titles) In some cases, the title of a news article may be as informative as the body text itself. For example, there is a category of news article known as clickbait in which the headline of the article is written in such a way that persuade the reader to click on the link. The authors of this specific kind of articles usually try to use some persuasive words to tempt readers to follow their articles [2, 5].

The study of news article titles could be very interesting in situations in which the webpage doesn't exists anymore but the tweet/post or a shared link still includes the title of the article. Moreover, if we could leverage the titles for classification of news articles, it enables us to predict trustworthiness of the content and help us to prevent browsing malicious webpages just by checking the titles. Having this in mind, the main goal here is to investigate how a model created out of the words (terms) of

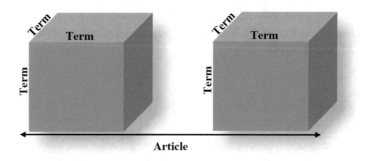

Fig. 3 Modeling content based information of articles using a 4 mode Tensor (3TA)

---

[1] We experimented with small values of that window and results were qualitatively similar.

the title can capture nuance differences between categories of fake news. To achieve this, we create a TTA Tensor but this time we just use titles' words (terms).

Step 2: k-NN Graph of News Articles

Decomposition of the tensor-based models we introduced in Step 1 provide us with three or four factor matrices (3 for TTA and 4 for 3TA) which are compact representation of each mode of the decomposed tensor (terms, terms or news articles) and each of which comprises concise set of latent patterns corresponding to that mode. Using the factor matrix corresponding to article mode (factor matrix $\mathbf{C}$ for TTA model and $\mathbf{D}$ for 3TA model), we can create a graphical representation of news articles in which each row (article) of the factor matrix $\mathbf{C}$ or $\mathbf{D}$ corresponds to a node in k-NN graph G. In other words, factor matrix $\mathbf{C}$ or $\mathbf{D}$ which is a representation of the news articles in the latent topic space can be used to construct a k-NN graph to find similar articles. To this end, we consider each row in $\mathbf{C}$ or $\mathbf{D} \in \mathbb{R}^{M \times R}$ as a point in R-dimensional space. We then compute ℓ_2 distance among nodes (news) to find the k-closest points for each data point in $\mathbf{C}$ or $\mathbf{D}$. In practice, the number of news articles is extremely large, so, in order to find the k-nearest-neighbors of each article efficiently, we propose to leverage a well-known kd-tree based optimizations in [22].

Each node in G represents a news article and each edge shows that two articles are similar in the embedding space. In this step, we only leverage the distance as a means to measure similarity between news articles, without much concern for the actual order of proximity. Thus, we enforce symmetry in the neighborhood relations, that is, if n_1 is a k-nearest-neighbor of news n_2, the opposite should also hold. The resultant graph G is an undirected, symmetric graph where each node is connected to at least k nodes. The graph can be compactly represented as an $M \times M$ adjacency matrix.

Step 3: Belief Propagation

Using the graphical representation of the news articles above, and considering that for a small set of those news articles we have ground truth labels, our problem becomes an instance of semi-supervised learning over graphs. We use a belief propagation algorithm which assumes homophily, because news articles that are connected in the k-NN graph are likely to be of the same type due to the construction method of the tensor embeddings; moreover, [14] demonstrates that such embeddings produces fairly homogeneous article groups. More specifically, we use the fast and linearized FaBP variant proposed in [17]. The algorithm is demonstrated to be insensitive to the magnitude of the known labels; thus, we consider that FaBP can achieve good learning performance only using a small number of known labels. Hence, we classify fake news articles in a semi-supervised fashion.

4 Implementation

In this section, we are going to describe the implementation details of our proposed method and datasets we used for experimenting.

First of all, we implemented our proposed method in MATLAB using the Tensor Toolbox [1] and for FaBP we used the implementation exists in [17]. For each experiment, we ran for 100 times and reported the average and standard deviation of the results. The reminder of this section describes the datasets we experimented on.

4.1 Dataset Description

For evaluation of our proposed method, we used two public datasets each of which consist of hundreds of articles and we created our own dataset that comprises more than 63 k articles, as shown in Table 1.

Public datasets The first public dataset i.e., *Dataset1* provided by [13] consists of 150 political news articles and is balanced to have 75 articles of each class. *Dataset2*, the second public dataset, provided by [11], comprises of 68 real and 69 fake news articles.

Our dataset For our dataset, we implemented a crawler in Python to crawl Twitter to collect news article URLs mentioned in some tweets during a 3-month period from June 2017 to August 2017. Our crawler extracts news content using the web API boilerpipe[2] and the Python library Newspaper3k.[3] For some few cases where these tools were not able to extract news content, we used Diffbot[4] which is another API to extract article text from web pages.

Table 1 Dataset specifics

Datasets	# fake news	# real news	# total
Dataset1 (Political)	75	75	150
Dataset2 (Bulgarian)	69	68	137
Our dataset	31,739	31,739	63,478

---

[2]http://boilerpipe-web.appspot.com/

[3]http://newspaper.readthedocs.io/en/latest/

[4]https://www.diffbot.com/dev/docs/article/

Table 2 Domain categories collected from BSDetector [4], as indicated in our dataset. The category descriptions are taken from [4]

Category	Description
Bias	"Sources that traffic in political propaganda and gross distortions of fact"
Clickbait	"Sources that are aimed at generating online advertising revenue and rely on sensationalist headlines or eye-catching pictures"
Conspiracy	"Sources that are well-known promoters of kooky conspiracy theories"
Fake	Sources that fabricate stories out of whole cloth with the intent of pranking the public
Hate	"Sources that actively promote racism, misogyny, homophobia, and other forms of discrimination"
Junk Science	"Sources that promote pseudoscience, metaphysics, naturalistic fallacies, and other scientifically dubious claims"
Rumor	"Sources that traffic in rumors, innuendo, and unverified claims"
Satire	"Sources that provide humorous commentary on current events in the form of fake news"

All real news articles were taken from Alexa[5] from 367 different domains, and fake news articles belong to 367 other domains identified by BSDetector (a crowd source tool box in form of browser extension to identify fake news sites) [4]. Table 2 demonstrates different categories specified by BSDetector. For this work, we considered news from all of these categories as fake news. With this in mind, our new dataset comprises 31,739 fake news and 409,076 real news articles. We randomly down-sampled the real class to get a create a balanced sample dataset. So, we created a sample balanced dataset consists of 31,739 articles of each class. The distribution of different fake categories has been illustrated in Fig. 4.

It is worth mentioning, we removed stopwords and punctuations from both body and the title of news articles and for all three datasets we preprocessed the data using tokenization and stemming.

5   Evaluation

In this section, we first introduce the evaluation metrics we leveraged for evaluating our proposed method and then we will discuss about experimental results of our basic tensor-based model i.e., model 1 (TTA) against some state-of-the-art baselines and will show how our method outperforms those baselines and then we will compare our model 2 (3TA) against model 1 (TTA) to show which one of these models performs better in terms of classification of misinformation. Finally, we will state the experimental results of model 3 (TTA out of titles) to see how successful is the TTA model when the only available information about the articles is the title.

---

[5]https://www.alexa.com/

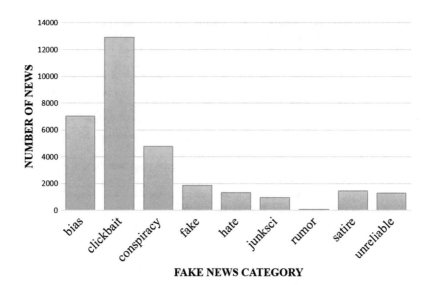

Fig. 4 Distribution of misinformation per domain category in our collected dataset

5.1 Experimental Results

Evaluation of basic tensor based model (TTA) against baselines In this section, we will discuss the experimental evaluation of using model 1 or TTA on body text of the articles. For this section, we used cross-validation where we evaluated different settings with respect to R i.e., decomposition rank and k i.e., the number of nearest neighbors which controls the density of the k-NN graph G.

Practically, decomposition rank is often set to be low for time and space reasons [26], so, we grid searched for values of R in range 1 to 20 and values of k in range 1 to 100. Actually, by increasing k we traded off the greater bias for less variance.

As a result of this experiment, parameters R and k both were set to be 10. As illustrated in Fig. 5, performance for values of k and R greater than 10, are very close. Moreover, using a small k value (for example, 1 or 2), leads to a poor accuracy because building a k-NN graph with small k results in a highly sparse graph which means limited propagation capacity for FaBP step. For all experiments, the accuracy is reported in terms of "unknown" or unspecified labeled articles in the propagation step.

As mentioned in Sect. 3, we consider two different tensor embeddings: *frequency-based* and *binary-based*. Figure 6 shows the performance of our proposed method using these two different tensors. As reported, the binary-based tensor performs better than frequency-based tensor in classification task. Thus, we used binary-based representations in evaluation. Later on, in Sect. 6, we will discuss intuition for why binary-based tensor results in better performance.

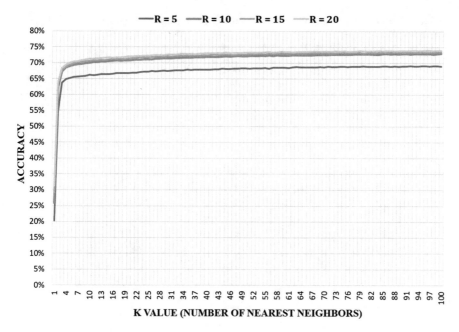

Fig. 5 Performance using different parameter settings for decomposition rank (R) and number of nearest neighbors (k)

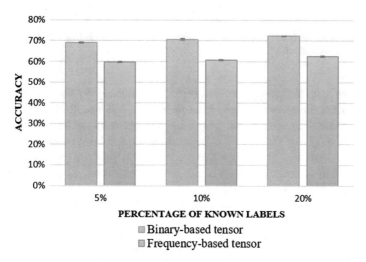

Fig. 6 Detection performance using different tensor representations

Evaluation of our method with different percentages p of known labels in range 5–30% is stated in Table 3.

As demonstrated, using only 10% of labeled articles we achieve an accuracy of 70.76%. To compare the robustness of our tensor based embedding against

Table 3 Performance of the proposed method using our dataset with different percentages of labeled news

%Labels	Accuracy	Precision	Recall	fF1
5%	69.12 ± 0.0026	69.09 ± 0.0043	69.24 ± 0.0090	69.16 ± 0.0036
10%	70.76 ± 0.0027	70.59 ± 0.0029	71.13 ± 0.0101	70.85 ± 0.0043
20%	72.39 ± 0.0013	71.95 ± 0.0017	73.32 ± 0.0043	72.63 ± 0.0017
30%	73.44 ± 0.0008	73.13 ± 0.0028	74.14 ± 0.0034	73.63 ± 0.0007

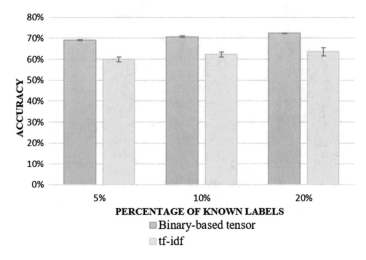

Fig. 7 Comparing accuracy of tensor-based embedding approach against $tf\text{-}idf$ approach for modeling of news content

widely used term frequency inverse-document-frequency ($tf\text{-}idf$) representation we constructed a k-NN graph built from the ($tf\text{-}idf$) representation of the articles as well. Figure 7 illustrates that tensor embeddings consistently results to better accuracy than ($tf\text{-}idf$) baseline for different known label percentages. This results empirically justify that binary-based tensor representations can capture spatial/contextual nuances of news articles better than widely used bag of words method for modeling the content of news articles.

We also experimented on an extremely sparse known labels setting i.e., known labels <5% for different values of nearest neighbors. Based on what shown in Fig. 8, our proposed method achieves an accuracy of 70.92% using just 2% of known labels when the number of nearest neighbors is 200. Indeed, the performance of our approach decreases fairly with even smaller proportions of known labels.

We also applied our proposed method on *Dataset1* and *Dataset2* and compared the resulted accuracy against the accuracy achieved by the following approaches:

- *SVM on content-based features* as proposed in [13]. We used suggested features extracted from news content and applied a SVM classifier and examined the performance using different percentages of training data.

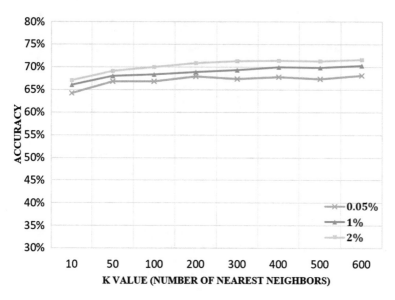

Fig. 8 Performance of proposed approach using extremely sparse (<5%) labeled set of articles and varying number of nearest neighbors

- *Logistic regression on content-based features* proposed by [11]. We used publicly available implementation of this work to extract linguistic (*n*-gram) features of the articles content.

Figure 9 shows experimental results for different approaches on *Dataset1*. As shown, our approach achieves better accuracy even with fewer labels. For instance, using only 30% of news labels we achieved 75.43% accuracy, whereas SVM(30%/70% train/test), SVM(5-fold cross-validation), and logistic regression (30%/70% train/test) attained 67.43%, 71% and 50.09% of accuracy, respectively. The accuracy achieved by SVM (5-fold cross-validation) was reported by Horne et al. in [13].

Moreover, we applied logistic regression and SVM, using 10%/90% train/test split on *Dataset2*. The accuracy of these approaches is equal to 59.84% and 64.79%, respectively whereas, our proposed method accuracy is equal to 67.38%.

One justification for reported improvements in terms of classification accuracy is that the tensor based modeling of news articles' content equips us with a means, which is potentially more capable of capturing nuanced patterns hidden within news content than widely used bag of words and tf-idf approaches.

Furthermore, another justification for having better performance in terms of accuracy even when we use small amount of labeled news articles is the fact that we leverage the k-NN graph in addition to belief propagation approach which allows us to exploit similarity between even unlabeled news articles and make our proposed approach stronger than supervised classification approaches when we experiment on extremely sparse known label regimes.

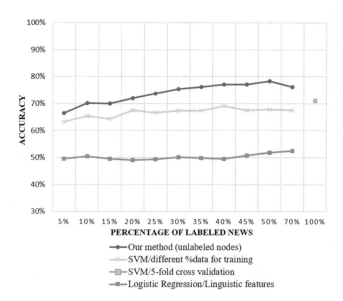

Fig. 9 Performance using *Dataset1* provided by Horne et al. [13]

Evaluation and comparison of text body constructed models i.e., model 1 (TTA) and model 2 (3TA) For the second experiment, we are going to investigate the effect of increasing number of co-occurred terms, which corresponds to a higher mode tensor based model, to evaluate whether or not considering a larger subset of co-occurred words provides us with a robuster model. For this reason, we created a 5 mode tensor-based model as we described in Sect. 3. We also used binary values for both models based on what we observed in previous experiment. For finding the best rank R of decomposition and the best number of nearest neighbors K, we again grid searched and based on the evaluation results we decided to set the R and K to 10 and 8 respectively. In Fig. 10, the classification performance of model 2 (3TA) in comparison to model 1 (TTA) for some ranks of decomposition and in terms of F1 score, precision and recall and accuracy has been demonstrated.

As shown in Fig. 10, increasing the number of co-occurred terms resulted in a considerable decline in the performance of the classification task. One possible justification is that by increasing the number of entries in co-occurrence tuple, we create a model that is more representative of an individual news article than a class of them. In other words, this model is essentially overfitting to specific articles. In fact, it is harder to find the nodes (articles) that share the same patterns of triple-way co-occurrence than a co-occurring pair. Thus, *our embedding of choice is model 1 (TTA)*.

Evaluation of titles constructed model i.e. model 3 (TTA out of titles) In this section, we will discuss the experimental results of model 3 or creating a TTA model out of articles' title. Since binary-based tensor leads to better results, as shown in model 1 evaluation, and due to the fact that the number of words within titles are

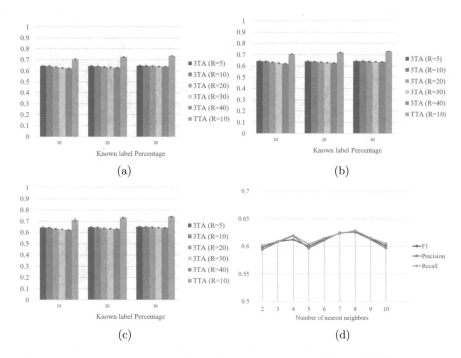

Fig. 10 F1 score, precision and recall for applying our proposed method on TTA and 3TA models and the effect of using different K on these metrics. As illustrated, TTA with chosen rank 10 outperforms 3TA with four different ranks of decomposition. Moreover, increasing the number of neighbors causes a significant decline in classification metrics. (**a**) F1-score. (**b**) Precision. (**c**) Recall. (**d**) Effect of using different values of K

quite less than body text, we chose binary-based tensor over frequency-based one for this experiment. We created 9 balanced binary-based tensors separately, each of which consists of 50% of under study category and 50% of real articles. To create balanced tensors, we used all the articles we have in our dataset for each category and added same number of real articles. Again we grid-searched for selecting the best hyperparameters and based on the search, we set the values of k-NN, window size and the R to 10, 5 and 10 respectively. The F1 score, precision, recall and accuracy for applying our proposed method on a TTA tensor made out of terms of the titles is shown in Fig. 11.

As demonstrated in Fig. 11, the classification results when using the only titles differs from category to category. Based on evaluation results of this experiment, we observe that the title of articles belong to news articles from clickbait, bias, rumor, hate and junk science categories are more informative and possibly convey more information about the content of the article. Moreover, since we are more successful in classification task for these categories, we may conclude that there are more meaningful co-occurrence for the title's terms of these categories in comparison to the rest of the misinformation categories, for which the title is not as informative as the entire content.

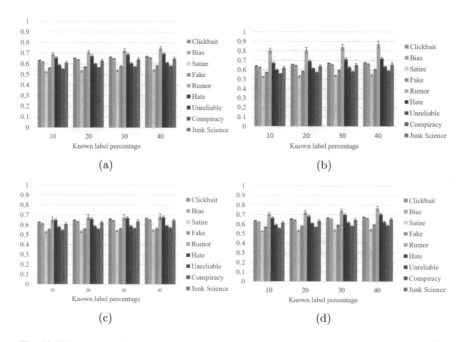

Fig. 11 F1 score, precision, recall and accuracy for applying our proposed method on a TTA tensor constructed out of terms of articles' title for different categories of fake articles. As shown, the results differ from category to category and best results belongs to clickbait, bias, rumor, hate and junk science categories. (**a**) F1-score. (**b**) Precision. (**c**) Recall. (**d**) Accuracy

6 Sensitivity Analysis

A question that may come to mind regarding the effectiveness of our proposed method is that how the length and categorical distribution of the articles impact the performance of our proposed method when we use frequency-based and binary-based tensor embeddings. To this end, we created sub-sampled datasets from our dataset proposed before, which meet the following conditions:

- News articles have similar content length and are selected across news categories
- News articles vary in length and belong to the same news category
- News articles have similar content length and belong to the same news category

We first evaluated our proposed method using a dataset where news articles have similar length and fake articles are selected across news categories. (Table 4 shows summary statistics for article length across each fake news category).

Figure 12 shows the accuracy achieved by our method using both binary-based and frequency-based tensor embeddings. The results suggest that performance is not sensitive to news category especially when length is standardized.

In addition, we evaluated our method using 8 sample datasets, one for each misinforming news category: *bias, clickbait, conspiracy, fake, hate, junk science,*

Table 4 Dataset statistics per fake news category

Dataset	Article length		
	Minimum	Mean	Maximum
Bias	18	363	5,903
Clickbait	18	355	10,955
Conspiracy	19	422	10,716
Fake	20	378	8,803
Hate	20	315	5,390
Junk Science	22	364	5,390
Satire	18	307	8,913
Unreliable	20	360	5,268

Fig. 12 Comparing performance of using binary vs. frequency-based tensor embeddings on news articles of all types with similar content length

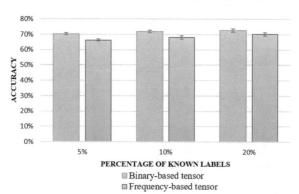

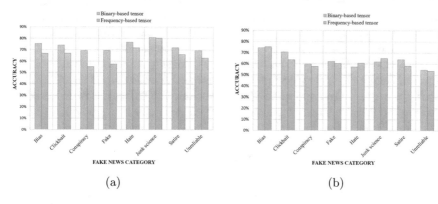

(a)   (b)

Fig. 13 Performance using binary vs. frequency-based tensor embeddings on category-partitioned news articles. (**a**) Varying in length. (**b**) With similar length

satire, and *unreliable*. Each dataset was balanced, containing the same number of fake and real articles. Note that in these sample datasets, news article length varies (see Table 4).

In Fig. 13, the accuracy achieved by our proposed method for each fake news category using both binary-based and frequency-based embeddings has been illustrated. These results show that the binary-based representation noticeably outperforms the frequency-based for all categories of fake news. We then performed a comparable experiment, but this time we only selected news articles that had almost the same length per category. Figure 13 demonstrates that the results of experiment are less clear – in fact, the results using both representations are quantitatively quite similar. These findings indicate that the news articles length greatly affects how well the different co-occurrence tensor embeddings perform. More specifically, we can conclude that binary-based tensors indicating boolean co-occurrence between words better captures spatial/contextual nuances of news articles that vary in length. On the same note, detection performance is relatively comparable for embedding types when considering articles of a fixed or almost-fixed length.

7   Related Work

7.1   Supervised Models

The majority of works have been proposed so far for misinformation detection task leverage supervised learning models based on extracted features from news content, user or contextual information. For instance, in [11], the authors proposed to extract linguistic (n-gram), credibility (punctuation, pronoun use, capitalization) and semantic features from the news content and then applied a logistic regression classifier to detect misinformation. In another work, authors used a SVM classifier on stylistic, complexity and psychological features extracted from content of the articles and classified them into real, fake and satirical news[13]. In [23], the authors leverage a naive-Bayes classifier on content, network and microblog-specific features for detecting rumors. For rumors detection task there is another work in which a Dynamic Series-Time Structure (DSTS) model is proposed to capture the social context of an event from content, user and propagation-based features[20]. Ma et al. [19] and Ruchansky et al. [25] model temporal structure using a recurrent neural network (RNN) to represent text and user characteristics. The majority of aforementioned works are based on complicated models and require human experts for feature based modeling and considerable volume of data for training and testing steps in a supervised manner, whereas we proposed a semi-supervised approach in which we construct a tensor based model out of words' co-occurrence that is not only simple and fast but also outperforms some state of the art supervised approaches by leveraging very few amount of labeled data.

7.2   Propagation Models

There are some other propagation-based models proposed for evaluating news credibility. For instance, the authors in [9] leverage a PageRank-like credibility propagation method on multi-typed network of events, tweets and users. For news verification task, the authors in [16] propose a credibility network based on positive and negative view points about news articles. In [15], a hierarchical propagation model on a three-layer credibility network is proposed which comprises event, sub-event and message layers. All of mentioned works require some kind of initial credibility values which obtained from the output of a supervised classifier. In contrast to these works, we leverage a semi-supervised propagation based which requires very few labels and performs admissibly even with only 5% of labeled data.

7.3   Ensemble Models

Fake news detection approaches mainly focus on one or two aspects of the data specially content of the articles[6, 33], user features [34], and temporal properties [18] which are widely used for misinformation detection. There are some approaches proposed recently which exploit multiple aspects of news articles to create a comprehensive model to investigate the trustworthiness of news articles. For example, in [29] the authors proposed an ensemble model created out of content-based information, user-user interactions, user-article interactions, and publisher-article relations. In [30], Shu et al. construct a joint model using news contents and user comments for fake news detection task.

Since our proposed method outperforms many state of the art content-base approaches for fake news detection task, we reserve further investigation of ensemble modeling using our tensor-based approach for future work.

8   Conclusions

In this work, we proposed a tensor-based semi-supervised approach for distinguishing misinformation. We proposed three different tensor-based models which decomposition of these tensors provide us with the patterns hidden amongst different categories of news articles. We leveraged a k-nearest neighbor graph to represent the proximity of news articles using the latent patterns extracted from decomposition of tensor models and apply belief propagation algorithm for propagating very few labels we have, throughout the k-nearest neighbor graph. We evaluate our proposed method on two public datasets and a dataset we created from over 63K articles. Experimental results on these real-world datasets illustrates that our approach outperform many state of the art content based methods in terms of accuracy even when we use very few known labels. More specifically, our method achieves accuracy of 75% on first public dataset and accuracy of 67% on second public dataset using 30% and 10% of the labels respectively. Moreover, the classification accuracy of our proposed method on our dataset is 71% using only 2%

of labels. Meanwhile, our tensor-based model created out of just titles of the articles is able to classify articles into different categories of fake news specially clickbait, bias, rumor, hate and junk science categories with accuracy of roughly 70% using just 30% of the labeled data.

Acknowledgements This research was supported by a gift from Snap Inc, an Adobe Data Science Faculty Award, by the Department of the Navy, Naval Engineering Education Consortium under award no. N00174-17-1-0005, and by the National Science Foundation CDS&E Grant no. OAC-1808591. Any opinions, findings, and conclusions or recommendations expressed here are those of the author(s) and do not necessarily reflect the views of the funding parties. We would also like to thank Daniel Fonseca for proofreading of the book chapter.

References

1. Bader, B.W., Kolda, T.G.: Matlab tensor toolbox version 2.6. Available online (2015)
2. Biyani, P., Tsioutsiouliklis, K., Blackmer, J.: "8 amazing secrets for getting more clicks": detecting clickbaits in news streams using article informality. In: Proceedings of the Thirtieth AAAI Conference on Artificial (AAAI'16), pp. 94–100 (2016)
3. Braunstein, A., Mézard, M., Zecchina, R.: Survey propagation: an algorithm for satisfiability. Random Struct. Algorithms **27**(2), 201–226 (2005). https://doi.org/10.1002/rsa.v27:2
4. BS Detector (2017). http://bsdetector.tech/
5. Chen, Y., Conroy, N., Rubin, V.: Misleading online content: recognizing clickbait as "false news" (2015). https://doi.org/10.1145/2823465.2823467
6. Guacho, G.B., Abdali, S., Shah, N., Papalexakis, E.E.: Semi-supervised content-based detection of misinformation via tensor embeddings. In: 2018 IEEE/ACM International Conference on Advances in Social Networks Analysis and Mining (ASONAM), pp. 322–325 (2018). https://doi.org/10.1109/ASONAM.2018.8508241
7. Gupta, A., Lamba, H., Kumaraguru, P.: $1.00 per rt #bostonmarathon #prayforboston: analyzing fake content on twitter. In: 2013 APWG eCrime Researchers Summit, pp. 1–12 (2013). https://doi.org/10.1109/eCRS.2013.6805772
8. Gupta, A., Lamba, H., Kumaraguru, P., Joshi, A.: Faking sandy: characterizing and identifying fake images on twitter during hurricane sandy. In: Proceedings of the 22Nd International Conference on World Wide Web, WWW '13 Companion, pp. 729–736. ACM, New York (2013). https://doi.org/10.1145/2487788.2488033
9. Gupta, M., Zhao, P., Han, J.: Evaluating Event Credibility on Twitter, pp. 153–164. https://doi.org/10.1137/1.9781611972825.14, http://epubs.siam.org/doi/abs/10.1137/1.9781611972825.14
10. Han, J., Kamber, M., Pei, J.: Data Mining: Concepts and Techniques, 3rd edn. Morgan Kaufmann Publishers Inc., San Francisco (2011)
11. Hardalov, M., Koychev, I., Nakov, P.: In Search of Credible News. Artificial Intelligence: Methodology, Systems, and Applications. AIMSA 2016 Lecture Notes in Computer Science, pp. 172–180 (2016). https://doi.org/10.1007/978-3-319-44748-3_17
12. Harshman, R.A.: Foundations of the PARAFAC procedure: models and conditions for an" explanatory" multi-modal factor analysis. UCLA Working Papers in Phonetics **16**(1), 84 (1970)
13. Horne, B.D., Adali, S.: This just in: fake news packs a lot in title, uses simpler, repetitive content in text body, more similar to satire than real news. CoRR abs/1703.09398 (2017). http://arxiv.org/abs/1703.09398
14. Hosseinimotlagh, S., Papalexakis, E.E.: Unsupervised content-based identification of fake news articles with tensor decomposition ensembles (2017)

15. Jin, Z., Cao, J., Jiang, Y.G., Zhang, Y.: News credibility evaluation on microblog with a hierarchical propagation model. In: 2014 IEEE International Conference on Data Mining, pp. 230–239 (2014). https://doi.org/10.1109/ICDM.2014.91
16. Jin, Z., Cao, J., Zhang, Y., Luo, J.: News verification by exploiting conflicting social viewpoints in microblogs (2016). https://www.aaai.org/ocs/index.php/AAAI/AAAI16/paper/view/12128
17. Koutra, D., Ke, T.Y., Kang, U., Chau, D., Pao, H.K., Faloutsos, C.: Unifying guilt-by-association approaches: theorems and fast algorithms. In: Machine Learning and Knowledge Discovery in Databases (ECML/PKDD), Lecture Notes in Computer Science, vol. 6912, pp. 245–260. Springer, Berlin/Heidelberg (2011)
18. Kumar, S., Shah, N.: False information on web and social media: a survey. arXiv preprint arXiv:1804.08559 (2018)
19. Ma, J., Gao, W., Mitra, P., Kwon, S., Jansen, B.J., Wong, K.F., Cha, M.: Detecting rumors from microblogs with recurrent neural networks. In: Proceedings of the Twenty-Fifth International Joint Conference on Artificial Intelligence (IJCAI'16), pp. 3818–3824. AAAI Press (2016). http://dl.acm.org/citation.cfm?id=3061053.3061153
20. Ma, J., Gao, W., Wei, Z., Lu, Y., Wong, K.F.: Detect rumors using time series of social context information on microblogging websites. In: Proceedings of the 24th ACM International on Conference on Information and Knowledge Management (CIKM '15), pp. 1751–1754. ACM, New York (2015). https://doi.org/10.1145/2806416.2806607
21. Papalexakis, E.E., Faloutsos, C., Sidiropoulos, N.D.: Tensors for data mining and data fusion: models, applications, and scalable algorithms. ACM Trans. Intell. Syst. Technol. 8(2), 16:1–16:44 (2016). https://doi.org/10.1145/2915921
22. Pelleg, D., Moore, A.: Accelerating exact k-means algorithms with geometric reasoning. In: Proceedings of the Fifth ACM SIGKDD International Conference on Knowledge Discovery and Data Mining, pp. 277–281. ACM (1999)
23. Qazvinian, V., Rosengren, E., Radev, D.R., Mei, Q.: Rumor has it: identifying misinformation in microblogs. In: Proceedings of the Conference on Empirical Methods in Natural Language Processing (EMNLP'11), pp. 1589–1599. Association for Computational Linguistics, Stroudsburg (2011). http://dl.acm.org/citation.cfm?id=2145432.2145602
24. Rubin, V.L., Conroy, N.J., Chen, Y., Cornwell, S.: Fake news or truth? using satirical cues to detect potentially misleading news (2016)
25. Ruchansky, N., Seo, S., Liu, Y.: CSI: a hybrid deep model for fake news. CoRR abs/1703.06959 (2017). http://arxiv.org/abs/1703.06959
26. Shah, N., Beutel, A., Gallagher, B., Faloutsos, C.: Spotting suspicious link behavior with fbox: an adversarial perspective. In: IEEE International Conference on Data Mining (ICDM), 2014, pp. 959–964. IEEE (2014)
27. Shu, K., Sliva, A., Wang, S., Tang, J., Liu, H.: Fake news detection on social media: a data mining perspective. CoRR abs/1708.01967 (2017). http://arxiv.org/abs/1708.01967
28. Shu, K., Le, T., Lee, D., Huan, L.: Deep headline generation for clickbait detection. In: IEEE International Conference on Data Mining (ICDM), pp. 467–476 (2018)
29. Shu, K., Sliva, A., Wang, S., Liu, H.: Beyond news contents: the role of social context for fake news detection. In: Proceedings of the Twelfth ACM International Conference on Web Search and Data Mining (WSDM'19), pp. 312–320 (2019)
30. Shu, K., Cui, L., Wang, S., Lee, D., Liu, H.: Defend: explainable fake news detection. In: Proceedings of 25th ACM SIGKDD International Conference on Knowledge Discovery and Data Mining (2019)
31. Sidiropoulos, N., De Lathauwer, L., Fu, X., Huang, K., Papalexakis, E., Faloutsos, C.: Tensor decomposition for signal processing and machine learning. IEEE Trans. Signal Process. **PP** (2016). https://doi.org/10.1109/TSP.2017.2690524
32. Silverman, C.: This analysis shows how fake election news stories outperformed real news on facebook. BuzzFeed News (2016)
33. Wu, L., Li, J., Hu, X., Liu, H.: Gleaning wisdom from the past: early detection of emerging rumors in social media. In: Proceedings of the 2017 SIAM International Conference on Data Mining, pp. 99–107. SIAM (2017)

34. Wu, L., Liu, H.: Tracing fake-news footprints: characterizing social media messages by how they propagate. In: Proceedings of the Eleventh ACM International Conference on Web Search and Data Mining, pp. 637–645. ACM (2018)
35. Yedidia, J., Freeman, W., Weiss, Y.: Constructing free-energy approximations and generalized belief propagation algorithms. IEEE Trans. Inf. Theory **51**, 2282–2312 (2005). https://doi.org/10.1109/TIT.2005.850085

Exploring the Role of Visual Content in Fake News Detection

Juan Cao, Peng Qi, Qiang Sheng, Tianyun Yang, Junbo Guo, and Jintao Li

Abstract The increasing popularity of social media promotes the proliferation of fake news, which has caused significant negative societal effects. Therefore, fake news detection on social media has recently become an emerging research area of great concern. With the development of multimedia technology, fake news attempts to utilize multimedia content with images or videos to attract and mislead consumers for rapid dissemination, which makes visual content an important part of fake news. Despite the importance of visual content, our understanding about the role of visual content in fake news detection is still limited. This chapter presents a comprehensive review of the visual content in fake news, including the basic concepts, effective visual features, representative detection methods and challenging issues of multimedia fake news detection. This chapter can help readers to understand the role of visual content in fake news detection, and effectively utilize visual content to assist in detecting multimedia fake news.

Keywords Fake news detection · Fake-news images · Social media · Image forensics · Image repurposing · Multimedia · Multi-modal · Deep learning · Computer vision

J. Cao (✉) · P. Qi · Q. Sheng · T. Yang
Key Laboratory of Intelligent Information Processing & Center for Advanced Computing Research, Institute of Computing Technology, CAS, Beijing, China

University of Chinese Academy of Sciences, Huairou, China
e-mail: caojuan@ict.ac.cn; qipeng@ict.ac.cn; shengqiang18z@ict.ac.cn; yangtianyun19z@ict.ac.cn

J. Guo · J. Li
Key Laboratory of Intelligent Information Processing & Center for Advanced Computing Research, Institute of Computing Technology, CAS, Beijing, China
e-mail: guojunbo@ict.ac.cn; jtli@ict.ac.cn

© Springer Nature Switzerland AG 2020
K. Shu et al. (eds.), *Disinformation, Misinformation, and Fake News in Social Media*, Lecture Notes in Social Networks,
https://doi.org/10.1007/978-3-030-42699-6_8

141

1 Introduction

1.1 Motivation

Social media platforms, such as Twitter[1] and Chinese Sina Weibo,[2] have become important access where people acquire the latest news and express their opinions freely.[3,4] However, the convenience and openness of social media have also promoted the proliferation of fake news, i.e., news with intentionally false information, which not only disturbed the cyberspace order but also caused many detrimental effects on real-world events. For example, in the political field, during the month before the 2016 U.S. presidential election campaign, the Americans encountered between one and three fake stories on average from known publishers[1], which inevitably misled the voters and influenced the election results; In the economic field, a piece of fake news claiming that Barack Obama was injured in an explosion wiped out $130 billion in stock value[5]; In the social field, dozens of innocent people were beaten to death by locals in India because of a piece of fake news about child trafficking that was widely spread on social media.[6] Hence, the automatic detection of fake news has become an urgent problem of great concern in recent years [18, 33, 49].

The development of multi-media technology promotes the evolution of self-media news from text-based posts to multimedia posts with images or videos, which attracts more attention from consumers and provides more credible storytelling. On the one hand, as a vivid description form, the visual content including images and videos is more attractive and salient than plain text and consequently boosts the news propagation. For instance, tweets with images get 18% more clicks, 89% more likes, and 150% more retweets than those without images.[7] On the other hand, visual content is often used as evidence of a story in our common sense, which can increase the credibility of the news.[8] Unfortunately, this advantage is also taken by fake news. For rapid dissemination, fake news usually contains misrepresented or even tampered images or videos to attract and mislead consumers. As a result, visual

---

[1] https://twitter.com/

[2] https://weibo.com/

[3] http://www.cac.gov.cn/2019-08/30/c_1124938750.htm

[4] https://www.journalism.org/2018/09/10/news-use-across-social-media-platforms-2018/

[5] https://www.telegraph.co.uk/finance/markets/10013768/Bogus-AP-tweet-about-explosion-at-the-White-House-wipes-billions-off-US-markets.html

[6] https://www.washingtonpost.com/world/asia_pacific/as-mob-lynchings-fueled-by-whatsapp-sweep-india-authorities-struggle-to-combat-fake-news/2018/07/02/683a1578-7bba-11e8-ac4e-421ef7165923_story.html

[7] https://www.invid-project.eu/tools-and-services/invid-verification-plugin/

[8] https://www.businesswire.com/news/home/20190204005613/en/Visual-SearchWins-Text-Consumers%E2%80%99-Trusted-Information

content has become an important part of fake news that cannot be neglected, making multimedia fake news detection a new challenge.

Multimedia fake news detection aims at effectively utilizing the information of several modalities, such as textual, visual and social modalities, to detect fake news. Visual modality can provide abundant visual information, which is preliminarily proven to be effective in fake news detection [15]. However, although the importance of exploiting visual content have been revealed, our understanding about the role of visual content in fake news detection remains limited. To further facilitate research on this problem, we present a comprehensive review of the visual content in fake news in this chapter, including the problem definition, available visual characteristics, representative detection approaches and challenging problems.

1.2 Problem Definition

In this subsection, we introduce the concept of fake news and analyze the different types of visual content in fake news.

Fake news is widely defined as news articles that are intentionally and verifiably false and could mislead consumers [1, 20, 33]. On the context of social multimedia, news articles refer to news posts with multimedia content that are published by users, so the general definition of fake news has been further refined [3, 5, 6, 46]. Formally, we state the refined definition as follows,

Definition 1.1 A piece of fake news is a news post that shares multimedia content that does not faithfully represent the event that it refers to.

In real-world scenarios, the visual content in fake news can be broadly classified into three categories: (1) visual content that is deliberately manipulated (also known as tampering, doctoring or photoshopping) or automatically generated by deep generative networks, which equals to *fake images/videos* in our common sense (see Fig. 1a), (2) visual content from an irrelevant event, such as a past event, a staged work or an artwork, that is reposted as being captured in the context of an emerging event (see Fig. 1b), or (3) visual content that is real (not edited) but is published together with a false claim about the depicted event (see Fig. 1c). All examples in Fig. 1 fall under our definition of fake news, because the images and associated texts jointly convey the misleading information regardless of the veracity of the textual or the visual content itself. For this reason, fake news is also referred to as *misleading content*[6] or *fauxtography*[46] in the context of social multimedia.

1.3 Organization

The remainder of this chapter is organized as follows. In Chap. 2, we introduce available visual features for fake news detection. We continue to present existing

Fig. 1 Examples of the visual content in fake news: (**a**) A tampered image where Putin is spliced on the middle seat at G-20 to show that he is in the center position of an intense discussion among other world leaders; (**b**) A real image captured in 2009 New York air crash, but it is claimed to be the wrecked Malaysia Airlines MH370 in 2014; (**c**) A real image taken at the moment when Hillary Clinton accidentally stumbled, but it was maliciously interpreted as evidence of Clinton's failing health

approaches utilizing visual content to detect fake news in Chap. 3. In Chap. 4, we discuss several challenging problems for multimedia fake news detection. Finally, we summarize available data repositories, tools (or software systems) and relevant competitions about multimedia fake news detection research in the appendix.

2   What Visual Content Tells?

Visual content has been shown as an important promoter for fake news propaganda.[9] At the same time, visual content also *tells* abundant cues for detecting fake news. To capture the distinctive characteristics of fake news, works extracted visual features from visual content (generally, images and videos), which can be categorized into four types: forensics features, semantic features, statistical features and context features.

2.1   Forensics Features

Since the addressed problem is the verification of multimedia posts, one reasonable approach would be to directly verify the truth of visual content, i.e., whether the image or video is captured in the event. Intuitively, if the visual content has undergone manipulation or severe re-compression, or is generated by deep learning techniques, the news post that it belongs to is likely to be fake. To access the authenticity, (blind) forensics features which can highlight the digitally edited

---

[9]https://www.wired.com/2016/12/photos-fuel-spread-fake-news/

traces of the visual content, are exploited in fake news detection from different perspectives, including the manipulation detection, generation detection and re-compression detection.

Manipulation Detection

Manipulation detection aims at looking for patterns or discontinuities left by operations such as splicing, copy-move and removal. The splicing refers to copying a part of one image and inserting it into another, while the copy-move and removal both happen in the same image. Because very few works [3] directly used these features in fake news detection yet, we also investigated the features mentioned in related works and summarized as follows:

- **Camera-related features** are particular patterns caused by the imaging pipeline, such as the sensor pattern noise and color filter array interpolation patterns, which can be destroyed by manipulation. In previous works, Photo-Response Non-Uniformity [11], noise inconsistencies [26] and local interpolation artifacts [10] were used to capture the change of those patterns.
- **Discontinuities in spatial features** are often left by forgery operations. To highlight these cues, gray-level run length features [47] and local binary patterns over the steerable-pyramid-transformed image [28] were exploited.

Note that some of them are only applicable to specific types of manipulation, which is unknown in practice. Also, some widely-spread manipulated images may have undergone multiple types of processing, increasing the challenge of capturing the traces of manipulation.

Generation Detection

As the rapid improvement of deep generative networks (especially generative adversarial network, GAN [12]), people can easily generate more photorealistic images and videos, making it hard to distinguish from natural ones. These misleading generated image and videos are often obtained by modifying the semantically-focused elements, for instance, the faces (mostly of celebrities), raising new threat to the trustworthiness of the visual content.

For generated fake images, existing works mostly focus on detecting with signal-level features. In the pixel domain, the co-occurrence matrices on three color channels were used for capturing spatial correlation characteristics, which were fed into the following convolutional neural network (CNN) for detection [29]. In contrast, McCloskey et al. started with the observation in the frequency domain that GAN images have more overlapping spectral responses among the RGB channels and negative weights than natural ones [27]. To represent these differences, this work introduced intensity noise histograms and over-/under- exposed rate.

For generated fake videos, most works are devoted to the detection of DeepFakes, a series of popular implementations for superimposing existing faces onto source videos. Works for DeepFakes detection mostly focused on the local features caused by the transformation in face-swapping such as the lacking of realistic eye blinking [22], the errors of 3D head poses introduced in face splicing for detection [44], and the artifacts left in warping to match the original faces [23].

Re-compression Detection

A fake image or video mostly suffers multiple compression in two situations: one is that the visual content is manipulated and re-saved at last, while the other is that it is repeatedly downloaded from and uploaded to the social media platform. These two situations probably indicate deliberate manipulation of visual content or misuse of the outdated, so we can detect fake news by predicting whether the attached visual content has been re-compressed.

For images, MediaEval VMU Task [3] (see in Appendix) extracted features directly related to the compression according to [2, 21], including probability map of the aligned/non-aligned double JPEG compression, potential primary quantization steps for the first 6 Discrete Cosine Transform (DCT) coefficients of the aligned/non-aligned double JPEG compression and block artifact grid. By thresholding the aligned/non-aligned JPEG compression maps above, Boididou et al. created two binary maps considered as object and background respectively and extracted descriptive statistics (maximum, minimum, mean, median, most frequent value, standard deviation and variance) for classification [4]. Qi et al. calculated block DCT coefficients and then performed Fourier Transform on them for enhancement to highlight the periodicity in the frequency domain caused by re-compression [30]. Furthermore, because multiple spreads may cause a dramatic decrease of clarity, no-reference quality measurement [41] can also indicate re-compression.

For videos, the methods exploited the presence of spikes in the Fourier transform of the energy of the displaced frame difference over time [37], blocking artifacts [24] and DCT coefficients of a macroblock [38] to detect the double-compression (mostly in MPEG videos).

2.2   Semantic Features

Fake news exploits the individual vulnerabilities of people and thus often relies on sensational or even fake images to provoke anger or other emotional response of consumers for promoting the spread of fake news. Thus, images in fake news often show some distinct characteristics in comparison with real news at the semantic level, such as visual impacts [16] and emotional provocations [33, 36] as Fig. 2

Fig. 2 Comparison of images in fake and real news images at the semantic level. We can find that fake news images are more visually striking and emotional provocative than real news images, even though they describe the same type of events such as fire (**a**), earthquake (**b**) and road collapse (**c**)

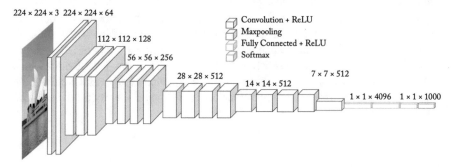

Fig. 3 Detailed architecture of the VGG16 framework

shows. Next, we introduce how to effectively extract semantic features of the visual content for fake news detection.

CNN has exhibited great power in understanding image semantics and obtaining corresponding feature representations, which can be used for various visual tasks. VGG [34] is one of the most popular CNN models, which is comprised of three basic types of layers: convolutional layers for extracting and transforming image features, pooling layers for reducing the parameters, and fully connected layers for classification tasks (see Fig. 3). Most of existing works based on multimedia content adopted the VGG model to extract visual semantic features for fake news detection [9, 15, 40].

In addition to the basic CNN, some recent works proposed novel CNN-based models to better capture the visual semantic characteristics of fake news. For example, Qi et al. proposed a multi-domain visual neural network (MVNN) to

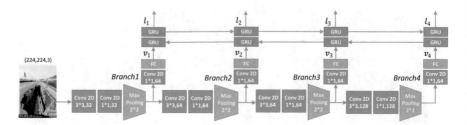

Fig. 4 Detailed architecture of the pixel domain sub-network in MVNN. For an input image, a multi-branch CNN-RNN network is utilized to extract and fuse its pixel-domain features of different semantic levels

fuse the visual information of frequency and pixel domains for detecting fake news, of which the pixel sub-network was used to extract visual semantic features (see Fig. 4) [30]. Specifically, two motivations were illustrated for the model design. First, CNN learns high-level semantic representations through layer-by-layer abstraction from local to the global view, while the low-level features will inevitably suffer some losses in the process of abstraction. Considering these semantic cues such as emotional provocations are related to many visual factors from low-level to high-level [19], a multi-branch CNN network was adopted to extract features of different semantic levels in the pixel sub-network. Second, there are strong bidirectional dependencies between different levels of features. For example, middle-level features such as textures, consist of low-level features such as lines, and meanwhile compose high-level features such as objects. Therefore, the sub-network also utilized the bidirectional GRU to model the relations from two different views.

2.3 Statistical Features

Visual content also has different distribution patterns between fake and real news on social media [17]. Intuitively, people tend to report the news with images taken by themselves at the event scene. If the event is real, then various images taken by different witnesses would be posted while if fake, there are many repeatedly posted images with almost the same content, just as Fig. 5 shows. Thus, we introduce visual statistical features to reflect this distributional difference between real and fake news.

Some works [17, 42, 43] used basic statistical features about the attached images to assist in fake news detection, usually from three aspects:

- **Count**: The occurrence number of images. For example, Wu et al. used the number of illustrations to assist detect fake news posts [42, 43], while Jin et al. used the ratio of news posts containing at least one or more than one images to the total posts in a news event to detect fake news events [17].

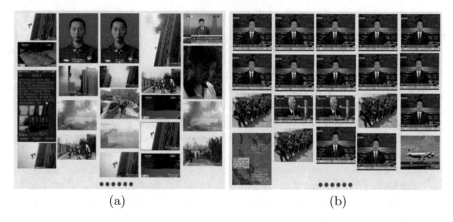

Fig. 5 Examples of images in the real and fake news event. Obviously, images in the real news event (**a**) are much more diverse than those in the fake one (**b**)

- **Popularity**: The number of sharing on social media, such as re-tweets and comments. Jin et al. defined the image with a high popularity as a hot image, and regarded the ratio of hot images to all distinct images in a news event as a statistical feature [17].
- **Type**: Some images have a particular type in resolution or style. For example, long images are images with a very large length-to-width ratio. The ratio of these types of images was also counted as a statistical feature [17].

In addition to these basic statistical features, Jin et al. also proposed five advanced statistical features as follows [17]:

- **Visual Clarity Score (VCS)**: Visual clarity score measures the distribution difference between two image sets: one is the image set in a certain news event (event set) and the other is the image set containing images from all events (collection set). This feature was defined as the Kullback-Leibler divergence between the two language model representing the event set and collection set, respectively. The bag-of-words image representation such as SIFT was used to define the language models for images. Specifically, the visual clarity score is

$$VCS = D_{KL}(p(w|c)\|p(w|k)), \tag{1}$$

where $p(w|c)$ and $p(w|k)$ denote the term frequency of visual word w in collection set and event set, respectively.
- **Visual Coherence Score (VCoS)**: Visual coherence score measures how coherent the images in a certain news event are. This feature is computed based on the visual similarity between any image pair within images in the target event image set, which is denoted as

$$VCoS = \frac{1}{|N(N-1)|} \sum_{i,j=1,\cdots,N;i\neq j} sim\left(x_i, x_j\right) \tag{2}$$

where N is number of the images in the event set, $sim\left(x_i, x_j\right)$ is the visual similarity between image x_i and image x_j. In implementation, the similarity between images is computed based on their GIST features.

- **Visual Similarity Distribution Histogram (VSDH)**: Visual similarity distribution histogram describes the image similarity distribution in a fine-granularity level, which is computed based on the whole similarity matrix of all images in a target news event. The similarity matrix S is quantified into an H-bin histogram by mapping each element in the matrix into its corresponding bin, which results in a feature vector of H dimensions representing the similarity relations among images,

$$VSDH(h) = \frac{1}{N^2} \left|\{(i,j)|i,j \leq N, m_{i,j} \in h - th \text{ bin}\}\right|, h = 1, \ldots, H \tag{3}$$

- **Visual Diversity Score (VDS)**: Visual diversity score measures the visual difference in the image set of a target news event. Assuming a ranking of images $x_1, x_2, \ldots, x_N$ in the event image set R, the diversity score of all images in R is,

$$VDS = \sum_{i=1}^{N} \frac{1}{i} \sum_{j=1}^{i} (1 - sim\left(x_i, x_j\right)) \tag{4}$$

In implementation, images are ranked according to their popularity on social media, based on the assumption that popular images may have better representation for the news event.

- **Visual Clustering Score (VCS)**: Visual clustering score evaluates the image distribution over all images in the news event from a clustering perspective. It was defined as the number of clusters formed by all images in a target news event. Hierarchical agglomerative clustering (HAC) algorithm is employed to cluster these images.

2.4 Context Features

According to our previous analysis, rumormongers usually use visual content from an irrelevant event to fabricate fake news. To make the fake news more reasonable, the selected visual content needs to be semantically coherent with the claim. Therefore, existing works about text-image semantic similarity aren't applicable for these manipulations. Instead, one of the most effective methods is to utilize the context information of visual content to fact-check whether the current event is the same as the original event it belongs to. Specifically, we introduce the following context features, which mainly extracted from two sources: the metadata of visual content and the external knowledge such as relevant web pages.

Metadata

Metadata is text information pertaining to an image/video file that is usually embedded into the file. Metadata includes not only the details relevant to the image/video itself such as file size but also the information about its production, such as position and time, which are often used in manually fact-checking [7, 45]. However, these features are not that helpful in practice because they usually become unavailable after default processing by social media.

External Knowledge

In addition to metadata, some works extracted context features from the external knowledge obtained through reverse image search. In contrast to classical image search, reverse image search takes an image as input and returns lo relevant web pages that include the corresponding image, title, description and time. This process can be easily automated and applied to a large number of images via some search engine APIs like google reverse image search.[10] Next, we introduce three context features as follows.

- **Timespan**: Timespan is defined as the time delay between the published time of the news and the earliest published time of the visual content. This feature is proposed to verify the originality of the visual content [35]. If the timespan is bigger than a specific threshold, then the visual content is probably from an irrelevant event.
- **Inter-claim similarity**: Inter-claim similarity is defined as the similarity between the claim and the textual contents of these crawled websites. Considering that the text information of these crawled websites is helpful for understanding the original event of the image, this feature is used to verify the event consistency between the textual claim and corresponding visual content [48].
- **Platform credibility**: Platform credibility means the credibility of the source platform where the visual content was published [48]. By using the dataset of Media Bias/Fact Check (MBFC),[11] a web site that provides factuality information about 2700+ media sources, each web page that is returned by the reverse image search was classified into the following categories: high factuality, low factuality and mixed factuality. The percentage of web pages from each category returned by the reverse image search was defined as the platform credibility feature.

---

[10]https://images.google.com/

[11]http://mediabiasfactcheck.com/

3   How Visual Content Helps?

In the previous section, we introduced four types of visual features from different perspectives, i.e., forensics features, semantic features, statistical features and context features, for multimedia fake news detection. These features reflect the characteristics of visual content and are usually combined in practice for covering more situations. In this section, we discuss the details of several existing approaches utilizing visual content to detect fake news, which can be broadly classified into content-based approaches and knowledge-based approaches. **Content-based approaches** focus on capturing and combining the cues from contents of different modalities for fake news detection, without using any reference datasets. **Knowledge-based approaches** aim to use external sources to fact-check input claims. They assume the existence of a relatively large reference dataset and assess the integrity of the news post by comparing it to one or more posts retrieved from the reference dataset.

3.1   Content-Based Approaches

A complete news story consists of textual and visual content simultaneously, which both provide distinctive cues for detecting fake news. Therefore, recent works on this problem focus on utilize and effectively fuse information from multiple modalities. Mostly, these works simply used a common recurrent neural network (RNN) and a pre-trained CNN to obtain the textual and visual semantic features. Next, we introduce three state-of-the-art approaches that fuse multimodal information for fake news detection.

Jin et al. [15] first incorporated multi-modal contents via deep neural networks to solve fake news detection problem. It proposed an innovative RNN with an attention mechanism (attRNN, see Fig. 6a) for effectively fusing the textual, visual and social context features. For a given tweet, its text and social context are first fused with an LSTM for a joint representation. This representation is then fused with visual features extracted from pre-trained deep CNN. The output of the LSTM at each time step is employed as the neuron-level attention to coordinate visual features during the fusion.

Wang et al. [40] proposed an end-to-end event adversarial neural network (EANN, see Fig. 6b) to detect newly-emerged fake news events based on event-invariant multi-modal features. It consists of three main components: the multi-modal feature extractor, the fake news detector, and the event discriminator. The multi-modal feature extractor is responsible for extracting the textual and visual features from posts. It cooperates with the fake news detector to learn the discriminable representation for fake news detection. The role of event discriminator is to remove the event-specific features and keep shared features among events.

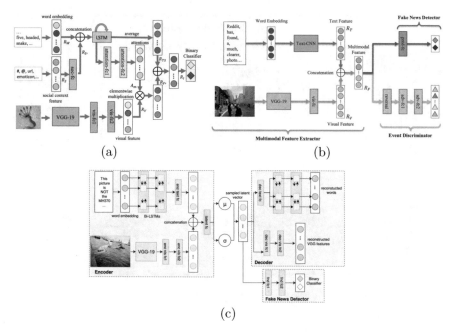

Fig. 6 Architectures of three state-of-the-art multi-modal models for fake news detection. (**a**) attRNN (**b**) EANN (**c**) MVAE

Dhruv et al. [9] utilized a multi-modal variational autoencoder (MVAE, see Fig. 6c) trained jointly with a fake news detector to learn a shared representation of textual and visual information. The model consists of three main components: an encoder, a decoder and a fake news detector module. The variational autoencoder is capable of learning probabilistic latent variable models by optimizing a bound on the marginal likelihood of the observed data. The fake news detector then utilizes the multi-modal representations obtained from the bi-modal variational autoencoder to classify posts as fake or not.

3.2 Knowledge-Based Approaches

Real-world multimedia news is often composed of multiple modalities, like the image or a video with associated text and metadata, where information about an event is incompletely captured by each modality separately. Such multimedia data packages, i.e., the tuples of multi-modal information of the posts, are prone to manipulations, where a subset of these modalities can be modified to misrepresent or repurpose the multimedia package. However, the details being manipulated are subtle and often interleaved with the truth, causing that the content-based approaches can hardly detect these manipulations. Faced with this problem, knowledge-based

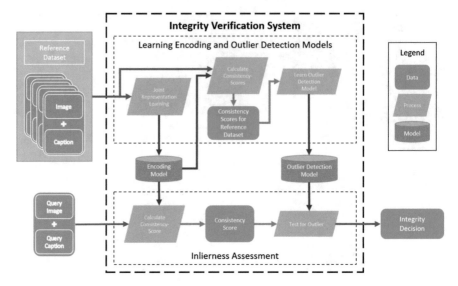

Fig. 7 The package integrity assessment system of [13]

approaches utilize external sources, a reference dataset of unmanipulated packages as a source of world knowledge, to help verify the semantic integrity of the multimedia news. In the following, we introduce some representative knowledge-based methods.

Jaiswal et al. [13] first formally defined the multimedia semantic integrity assessment problem and combined deep multi-modal representation learning with outlier detection methods to assess whether a caption was consistent with the image in its package (see Fig. 7). Data packages in the reference dataset were used to train a deep multi-modal representation learning model, which was then used to assess the integrity of query packages by calculating image-caption consistency scores and employing outlier detection models to find their inlierness with respect to the reference dataset.

Similarly, Sabir et al. [31] proposed a novel deep multi-modal model (see Fig. 8) to verify the integrity of multimedia packages. The proposed model consists of four modules: (1) feature extraction, (2) feature balancing, (3) package evaluation and (4) integrity assessment. For each query package, the model first uses similarity scoring to retrieve a package from the reference dataset, taking the query package and the top-1 related package as the input of the model. After passing to the feature extraction and balancing modules, query and retrieved packages are transformed into a single feature vector. The package evaluation module, the core of the proposed model, consists of the related package and single package sub-modules. The related package sub-module consisted of two siamese networks. The first network is a relationship classifier that verifies whether the query package and top-1 package are indeed related, while the second network is a manipulation detector that determines whether the query package is a manipulated version of the top-1 retrieved package.

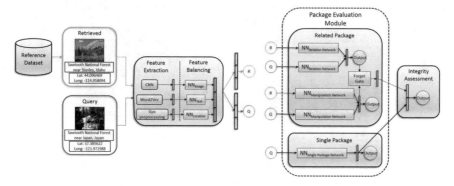

Fig. 8 The package integrity assessment model of [31]

Since manipulation detection is dependent on the relatedness of the two packages, the relationship classifier controls a forget gate which scales the feature vector of the manipulation detector according to the relatedness between the two packages. In the meantime, a single package module verifies the coherency (i.e., integrity) of the query package alone. The integrity assessment module concatenated feature vectors from both related and single package modules for manipulation classification.

One of the main challenges for developing multimedia semantic integrity assessment methods is the lack of training and evaluation data. In light of this, Jaiswal et al. [14] proposed a novel framework, Adversarial Image Repurposing Detection (AIRD) (see Fig. 9), for image repurposing detection, which can be trained in the absence of training data containing manipulated metadata. AIRD is to simulate the real-world adversarial interplay between a bad actor who repurposes images with counterfeit metadata and a watchdog who verifies the semantic consistency between images and their accompanying metadata. More specifically, AIRD consists of two models: a counterfeiter and a detector, which are trained in an adversarial way. While the detector gathers evidence from the reference set, the counterfeiter exploits it to conjure convincingly deceptive fake metadata for a given query package.

4 Challenging Problems

In the previous sections, we introduce several visual features and existing approaches based on visual content for effective fake news detection. Despite the research developments on the multimedia fake news detection problem, there are still some specific challenges that need to be considered.

One major challenge is the lacking of labeled data. Although the multimedia content is rapidly growing nowadays, datasets about multimedia fake news are scarce, which hinders the development of this research field. To tackle this challenge, on the one hand, we encourage researchers to pay more attention to

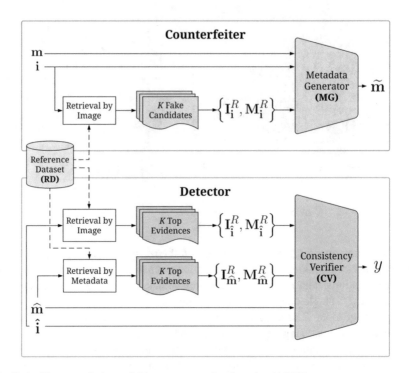

Fig. 9 Architecture of adversarial image repurposing detection (AIRD)

constructing and releasing high-quality labeled datasets. On the other hand, it is important to study multimedia fake news detection in a weakly supervised setting, i.e., with limited or no label data for training. For example, Jin et al. [16] constructs a large-scale weakly-labeled dataset as auxiliary to overcome the data scarcity issue, and proposes a domain transferred deep CNN to detect the fake news images.

Besides, another critical challenge is the explainability of fake news detection, i.e., why a model determines a particular piece of news as fake. Although computational detection of fake news has produced some promising results, the explainability of such detection remains largely unsolved, making the judgments unconvincing. In recent years, fact-checking approaches have aroused the attention of researchers, which could offer a new way to tackle this challenge. Different from traditional style-based fake news detection, these approaches utilize external resources (also known as knowledge) as evidence to fact-check a given piece of news is fake or real. For multimedia content, the relationship between the textual and visual content and metadata is a powerful clue, which can be combined with the external knowledge to make inferences. These approaches are helpful for better understanding and explaining the decision made by algorithms according to the involved evidence and visible inference process.

Acknowledgements This work was supported by the National Natural Science Foundation of China (U1703261).

Appendix

Data Repositories

The step above all to detect fake news is to collect a real-world benchmark dataset. Though several text-based fake news datasets [25, 39] have been released, publicized multimedia fake news datasets remain rare, hindering the development of fake multimedia news detection. We here introduce representative multimedia datasets in fake news detection as follows.

MediaEval-VMU[12]: The earliest publicly available multimedia verification corpus originates from the MediaEval 2015 Verifying Multimedia Use (VMU) task [3], which is further extended in 2016 [5]. In the latest version, the dataset consists of tweets from Twitter related to 17 events (or hoaxes) that comprise in total 193 cases of real images, 218 cases of misused (fake) images and two cases of misused videos, associated with 6,225 real and 9,404 fake tweets posted by 5,895 and 9,025 unique users, respectively.

TMM17: Due to the insufficiency of images in previous works like VMU, Jin et al. [17] collected a new dataset by crawling posts related to the authoritatively verified events from Weibo. The dataset is constituted of 146 news events with 50,287 posts posted by 42,441 distinct users. A total of 25,953 images are attached to 19,762 of the posts. Note that this work focuses on event-level detection, so there exist posts with no image attached.

MM17[15]: This multimedia dataset is especially for multi-modal fake news detection. The authors used similar sources as [17], but text-only posts and posts with duplicated, small-size and large-height images were removed. The dataset finally consists of 9,528 posts, with balanced amounts of fake and real news.

FakeNewsNet[13]: In [32], Shu et al. collected fake news articles instead of short statements by traversing the fact-check websites such as PolitiFact[14] and GossipCop[15] and then searching for the web pages of corresponding articles. Totally, 336 fake and 447 real news articles contain images in PolitiFact part, while 1,650 fake and 16,767 real do in GossipCop part.

MCG-FNeWS[16][8]: The first version of this dataset was released for the False News Detection Competition 2019. The data was collected from Weibo official

---

[12]https://github.com/MKLab-ITI/image-verification-corpus

[13]https://github.com/KaiDMML/FakeNewsNet

[14]https://www.politifact.com/

[15]https://www.gossipcop.com/

[16]http://mcg.ict.ac.cn/wordpress/share/mcg-fnews/

debunking center[17] and news verification system AI-Shiyao[18] and reorganized for different sub-tasks in the competition. For multi-modal detection sub-task, the whole set consists of 46,373 posts (23,186 real and 23,187 fake) with 41,937 images (24,794 in real posts and 17,170 in fake posts).

EMNLP19[19] [48]: This dataset is especially for verifying the claims about images. The image-related news was collected from two sources: A section of Snopes.com named *Fauxtography*[20] for all false image-related news and a small fraction of true news; Reuters' *Picture of the Year* from 2015 to 2018 for most of true news. In total, this dataset contains 592 true and 641 false image-claim pairs.

Tools

In addition to methods, tools to verify the visual content of fake news online is valuable due to its convenience to non-technical users. In this subsection, we introduce some publicly available tools for multimedia content verification.

Google Reverse Image Search: A service of searching by an image from Google. The verifiers may upload the image or input the image URL to find similar images as well as the web pages containing them. Other substitutions like Baidu Images,[21] provide similar service.

FotoForensics[22]: A website for forensics analysis of JPEG or PNG image, providing information including error level, hidden pixels, metadata and JPEG quality. Over 3.3 million images were analyzed by the service so far.

Image Verification Assistant[23]: A website to analyze the veracity of online media supported by REVEAL project. For an image, it extracts and visualizes the metadata and detects various types of forensics features, such as double JPEG quantization, JPEG Ghosts, JPEG blocking artifact, error level analysis, high-frequency noise and median filtering noise residue.

Fake Video News Debunker[24]: A free plugin that runs in Google Chrome or FireFox to verify videos and images. This integrated plugin provides service to obtain contextual information from Youtube or Facebook, extract keyframes for reverse image search, list the metadata and perform forensic analysis.

---

[17] https://service.account.weibo.com/

[18] https://www.newsverify.com/

[19] http://gitlab.com/didizlatkova/fake-image-detection

[20] https://www.snopes.com/fact-check/category/photos/

[21] https://image.baidu.com/

[22] http://www.fotoforensics.com/

[23] http://reveal-mklab.iti.gr/reveal/

[24] https://www.invid-project.eu/tools-and-services/invid-verification-plugin/

Relevant Competitions

To attract the attention from academia and industry and further promote the development of detection technology, considerable competitions for fake news detection were held but very few of them provided visual contents. Here, we introduce two competitions where visual contents can be exploited.

Verifying Multimedia Use (VMU): A part of the MediaEval Benchmark in 2015[3] and 2016[5], dealing with the automatic detection of manipulation and misuse of web multimedia content. A fake tweet was defined as a tweet that shared multimedia content inconsistent with the event it referred to. In 2015, participants were asked to predict the veracity (fake, real or unknown), given a tweet and the accompanying multimedia item (image or video) from an event. In 2016, a new related sub-task was added to detect image tampering.

False News Detection Competition 2019[25]: A competition held for false news detection on Weibo, with three sub-tasks: text-only, image-only and multi-modal detection. In image-only detection, models had to predict whether the image was attached to a false news post. In multi-modal detection, text, images and user profiles were all available to predict the veracity of the post.

References

1. Allcott, H., Gentzkow, M.: Social media and fake news in the 2016 election. J. Econ. Perspect. **31**(2), 211–236 (2017)
2. Bianchi, T., Piva, A.: Image forgery localization via block-grained analysis of jpeg artifacts. IEEE Trans. Inf. Forensic. Secur. **7**(3), 1003–1017 (2012)
3. Boididou, C., Andreadou, K., Papadopoulos, S., Dang-Nguyen, D.T., Boato, G., Riegler, M., Kompatsiaris, Y., et al.: Verifying multimedia use at mediaeval 2015. In: MediaEval (2015)
4. Boididou, C., Papadopoulos, S., Dang-Nguyen, D.T., Boato, G., Kompatsiaris, Y.: The certh-unitn participation@ verifying multimedia use 2015. In: MediaEval (2015)
5. Boididou, C., Papadopoulos, S., Dang-Nguyen, D.T., Boato, G., Riegler, M., Middleton, S.E., Petlund, A., Kompatsiaris, Y., et al.: Verifying multimedia use at mediaeval 2016. In: MediaEval (2016)
6. Boididou, C., Papadopoulos, S., Zampoglou, M., Apostolidis, L., Papadopoulou, O., Kompatsiaris, Y.: Detection and visualization of misleading content on twitter. Int. J. Multimed. Inf. Retr. **7**(1), 71–86 (2018)
7. Brandtzaeg, P.B., Lüders, M., Spangenberg, J., Rath-Wiggins, L., Følstad, A.: Emerging journalistic verification practices concerning social media. Journal. Pract. **10**(3), 323–342 (2016)
8. Cao, J., Sheng, Q., Qi, P., Zhong, L., Wang, Y., Zhang, X.: False news detection on social media. arXiv preprint arXiv:1908.10818 (2019)
9. Dhruv, K., Jaipal Singh, G., Manish, G., Vasudeva, V.: Mvae: multimodal variational autoencoder for fake news detection. In: Proceedings of the 2019 World Wide Web Conference. ACM (2019)

---

[25]https://www.biendata.com/competition/falsenews/

10. Ferrara, P., Bianchi, T., De Rosa, A., Piva, A.: Image forgery localization via fine-grained analysis of cfa artifacts. IEEE Trans. Inf. Forensic. Secur. **7**(5), 1566–1577 (2012)
11. Goljan, M., Fridrich, J., Chen, M.: Defending against fingerprint-copy attack in sensor-based camera identification. IEEE Trans. Inf. Forensic. Secur. **6**(1), 227–236 (2010)
12. Goodfellow, I., Pouget-Abadie, J., Mirza, M., Xu, B., Warde-Farley, D., Ozair, S., Courville, A., Bengio, Y.: Generative adversarial nets. In: Advances in Neural Information Processing Systems, pp. 2672–2680 (2014)
13. Jaiswal, A., Sabir, E., AbdAlmageed, W., Natarajan, P.: Multimedia semantic integrity assessment using joint embedding of images and text. In: Proceedings of the 25th ACM international conference on Multimedia, pp. 1465–1471. ACM (2017)
14. Jaiswal, A., Wu, Y., AbdAlmageed, W., Masi, I., Natarajan, P.: Aird: adversarial learning framework for image repurposing detection. In: Proceedings of the IEEE Conference on Computer Vision and Pattern Recognition, pp. 11330–11339 (2019)
15. Jin, Z., Cao, J., Guo, H., Zhang, Y., Luo, J.: Multimodal fusion with recurrent neural networks for rumor detection on microblogs. In: Proceedings of the 2017 ACM on Multimedia Conference, pp. 795–816. ACM (2017)
16. Jin, Z., Cao, J., Luo, J., Zhang, Y.: Image credibility analysis with effective domain transferred deep networks. arXiv preprint arXiv:1611.05328 (2016)
17. Jin, Z., Cao, J., Zhang, Y., Zhou, J., Tian, Q.: Novel visual and statistical image features for microblogs news verification. IEEE Trans. Multimed. **19**(3), 598–608 (2017)
18. Kumar, S., Shah, N.: False information on web and social media: a survey. arXiv preprint arXiv:1804.08559 (2018)
19. Lang, P.J.: A bio-informational theory of emotional imagery. Psychophysiology **16**(6), 495–512 (1979)
20. Lazer, D.M., Baum, M.A., Benkler, Y., Berinsky, A.J., Greenhill, K.M., Menczer, F., Metzger, M.J., Nyhan, B., Pennycook, G., Rothschild, D., et al.: The science of fake news. Science **359**(6380), 1094–1096 (2018)
21. Li, W., Yuan, Y., Yu, N.: Passive detection of doctored jpeg image via block artifact grid extraction. Signal Process. **89**(9), 1821–1829 (2009)
22. Li, Y., Chang, M.C., Lyu, S.: In Ictu Oculi: exposing ai generated fake face videos by detecting eye blinking. arXiv preprint arXiv:1806.02877 (2018)
23. Li, Y., Lyu, S.: Exposing deepfake videos by detecting face warping artifacts. In: Proceedings of the IEEE Conference on Computer Vision and Pattern Recognition Workshops, pp. 46–52 (2019)
24. Luo, W., Wu, M., Huang, J.: Mpeg recompression detection based on block artifacts. In: Security, Forensics, Steganography, and Watermarking of Multimedia Contents X, vol. 6819, p 68190X. International Society for Optics and Photonics (2008)
25. Ma, J., Gao, W., Mitra, P., Kwon, S., Jansen, B.J., Wong, K.F., Cha, M.: Detecting rumors from microblogs with recurrent neural networks. In: Proceedings of the Twenty-Fifth International Joint Conference on Artificial Intelligence, pp. 3818–3824. AAAI Press (2016)
26. Mahdian, B., Saic, S.: Using noise inconsistencies for blind image forensics. Image Vision Comput. **27**(10), 1497–1503 (2009)
27. McCloskey, S., Albright, M.: Detecting GAN-generated imagery using color cues. arXiv preprint arXiv:1812.08247 (2018)
28. Muhammad, G., Al-Hammadi, M.H., Hussain, M., Bebis, G.: Image forgery detection using steerable pyramid transform and local binary pattern. Mach. Vis. Appl. **25**(4), 985–995 (2014)
29. Nataraj, L., Mohammed, T.M., Manjunath, B., Chandrasekaran, S., Flenner, A., Bappy, J.H., Roy-Chowdhury, A.K.: Detecting GAN generated fake images using co-occurrence matrices. arXiv preprint arXiv:1903.06836 (2019)
30. Qi, P., Cao, J., Yang, T., Guo, J., Li, J.: Exploiting multi-domain visual information for fake news detection. In: 19th IEEE International Conference on Data Mining. IEEE (2019)
31. Sabir, E., AbdAlmageed, W., Wu, Y., Natarajan, P.: Deep multimodal image-repurposing detection. In: 2018 ACM Multimedia Conference on Multimedia Conference, pp. 1337–1345. ACM (2018)

32. Shu, K., Mahudeswaran, D., Wang, S., Lee, D., Liu, H.: Fakenewsnet: a data repository with news content, social context and dynamic information for studying fake news on social media. arXiv preprint arXiv:1809.01286 (2018)
33. Shu, K., Sliva, A., Wang, S., Tang, J., Liu, H.: Fake news detection on social media: a data mining perspective. ACM SIGKDD Explor. Newsletter **19**(1), 22–36 (2017)
34. Simonyan, K., Zisserman, A.: Very deep convolutional networks for large-scale image recognition. arXiv preprint arXiv:1409.1556 (2014)
35. Sun, S., Liu, H., He, J., Du, X.: Detecting event rumors on Sina Weibo automatically. In: Asia-Pacific Web Conference, pp. 120–131. Springer (2013)
36. Sunstein, C.R.: On Rumors: How Falsehoods Spread, Why We Believe Them, and What Can Be Done. Princeton University Press, Princeton (2014)
37. Wang, W., Farid, H.: Exposing digital forgeries in video by detecting double MPEG compression. In: Proceedings of the 8th Workshop on Multimedia and Security, pp. 37–47. ACM (2006)
38. Wang, W., Farid, H.: Exposing digital forgeries in video by detecting double quantization. In: Proceedings of the 11th ACM Workshop on Multimedia and Security, pp. 39–48. ACM (2009)
39. Wang, W.Y.: "Liar, liar pants on fire": a new benchmark dataset for fake news detection. In: Proceedings of the 55th Annual Meeting of the Association for Computational Linguistics (Volume 2: Short Papers), pp. 422–426 (2017)
40. Wang, Y., Ma, F., Jin, Z., Yuan, Y., Xun, G., Jha, K., Su, L., Gao, J.: Eann: event adversarial neural networks for multi-modal fake news detection. In: Proceedings of the 24th ACM SIGKDD International Conference on Knowledge Discovery & Data Mining, pp. 849–857. ACM (2018)
41. Wang, Z., Sheikh, H.R., Bovik, A.C.: No-reference perceptual quality assessment of jpeg compressed images. In: Proceedings. International Conference on Image Processing, vol. 1, pp. I–I. IEEE (2002)
42. Wu, K., Yang, S., Zhu, K.Q.: False rumors detection on Sina Weibo by propagation structures. In: 2015 IEEE 31st International Conference on Data Engineering, pp. 651–662. IEEE (2015)
43. Yang, F., Liu, Y., Yu, X., Yang, M.: Automatic detection of rumor on Sina Weibo. In: Proceedings of the ACM SIGKDD Workshop on Mining Data Semantics. p. 13. ACM (2012)
44. Yang, X., Li, Y., Lyu, S.: Exposing deep fakes using inconsistent head poses. In: ICASSP 2019–2019 IEEE International Conference on Acoustics, Speech and Signal Processing (ICASSP), pp. 8261–8265. IEEE (2019)
45. Zampoglou, M., Papadopoulos, S., Kompatsiaris, Y., Bouwmeester, R., Spangenberg, J.: Web and social media image forensics for news professionals. In: Tenth International AAAI Conference on Web and Social Media (2016)
46. Zhang, D.Y., Shang, L., Geng, B., Lai, S., Li, K., Zhu, H., Amin, M.T., Wang, D.: Fauxbuster: a content-free fauxtography detector using social media comments. In: 2018 IEEE International Conference on Big Data (Big Data), pp. 891–900. IEEE (2018)
47. Zhao, X., Li, J., Li, S., Wang, S.: Detecting digital image splicing in chroma spaces. In: International Workshop on Digital Watermarking, pp. 12–22. Springer (2010)
48. Zlatkova, D., Nakov, P., Koychev, I.: Fact-checking meets fauxtography: verifying claims about images. In: Proceedings of the 2019 Conference on Empirical Methods in Natural Language Processing and the 9th International Joint Conference on Natural Language Processing (EMNLP-IJCNLP), pp. 2099–2108 (2019)
49. Zubiaga, A., Aker, A., Bontcheva, K., Liakata, M., Procter, R.: Detection and resolution of rumours in social media: a survey. ACM Comput. Surv. (CSUR) **51**(2), 32 (2018)

Credibility-Based Fake News Detection

Niraj Sitaula, Chilukuri K. Mohan, Jennifer Grygiel, Xinyi Zhou,
and Reza Zafarani

Abstract Fake news can significantly misinform people who often rely on online sources and social media for their information. Current research on fake news detection has mostly focused on analyzing fake news content and how it propagates on a network of users. In this paper, we emphasize the detection of fake news by assessing its credibility. By analyzing public fake news data, we show that information on news sources (and authors) can be a strong indicator of credibility. Our findings suggest that an author's history of association with fake news, and the number of authors of a news article, can play a significant role in detecting fake news. Our approach can help improve traditional fake news detection methods, wherein content features are often used to detect fake news.

Keywords Fake news · Misinformation · Credibility assessment · Social media · Data analysis

1 Introduction

In this digital age, news and information are mostly received from various online sources. Surveys have shown that a large number of online users depend on social media for the news: 51% in Australia, 46% in Italy, 40% in the United States, and 36% in the United Kingdom [6]. Hence, fake news can misinform many people who rely on online news and/or social media for the information.

The impact of fake news has been widely discussed after the 2016 U.S. presidential election. A study by Silverman [30] shows that for the top 20 election stories in 2016, the top 20 fake news stories had 8,711,000 shares, reactions, and comments on Facebook. These user engagement numbers were significantly higher

N. Sitaula (✉) · C. K. Mohan · J. Grygiel · X. Zhou · R. Zafarani
Syracuse University, Syracuse, NY, USA
e-mail: nsitaula@syr.edu; mohan@syr.edu; jgrygiel@syr.edu; zhouxinyi@data.syr.edu;
reza@data.syr.edu

© Springer Nature Switzerland AG 2020   163
K. Shu et al. (eds.), *Disinformation, Misinformation, and Fake News in Social
Media*, Lecture Notes in Social Networks,
https://doi.org/10.1007/978-3-030-42699-6_9

than those for the top 20 real stories, with 7,367,000 shares, reactions, and comments on Facebook during the same time period. These concerns motivate assessing news credibility and detecting fake news *before* it spreads online.

Detecting fake news has gained attention from many academic researchers as well as other organizations. Although some fact checking websites exist, such as FactCheck[1] and PolitiFact,[2] the problem of detecting fake news is far from being solved. Manually verifying each and every fact in the news is extremely difficult with the high volume of data being created and shared every minute. Furthermore, it has become extremely difficult to decide whether a news article is fake or credible, since fake news articles often contain false information as well as some facts. Potthast, et al. [24], observe that fake news articles may contain facts, and credible news articles may contain factual errors. Hence, an automated process to detect fake news based only on content verification may not be effective. If we emphasize past information about the sources (authors or URLs) of news articles, then deception is still possible with new URLs and new fake author names. Our goal is to identify general indicators of news credibility, using both (1) *source* and (2) *content* perspectives to help detect fake news.

Using public data for fake news detection [27], we have analyzed multiple features of information related to the sources and contents of news articles. Our analyses demonstrate that fake news can be distinguished from true news based on features related to source and content. We also observe that while some features exhibit differences between fake and factual news, they do not help better predict fake news.

This paper focuses on finding signals or indicators of news credibility that can help detect fake news. Our findings suggest that the information about authors of news articles can indicate news credibility and help detect fake news. Using only information on the number of authors and the authors' publication history, classifiers were able to obtain >0.75 average F_1-score. When content related features were added to these features, we observed further improvements when detecting fake news.

In the following, we detail our analysis on various aspects of credibility. We review related work in Sect. 2, followed by a brief description of the dataset used for analysis in Sect. 3. Section 4 provides our analysis of credibility based on the source of the news and Sect. 5 details credibility factors based on the content of news. Based on our analysis on source and content credibility, we build predictive models to detect fake news, which we detail in Sect. 6. We conclude in Sect. 7 and present some directions for future work.

---

[1] https://www.factcheck.org

[2] https://www.politifact.com/

2   Related Work

We briefly discuss research on fake news detection and credibility assessment.

Fake News Research Fake news has been an active area of research, where is has been detected often by relying on (1) news content and/or (2) social context information. Often formed from text and images along with news sources (authors or websites), news content has been utilized in various ways to detect fake news. Text has been represented as a set of (subject, predicate, object) features and used to predict fake news by developing link prediction algorithms, i.e., how likely the extracted predicate connects the subject with the specific object [4, 10, 26]. Such textual information can be represented as style features at various language levels as well, e.g., lexicon-level [23, 32, 35, 36], syntax-level [5, 36], semantic-level [23], and discourse-level [15, 25], based on n-grams [23], Term Frequency–Inverse Document Frequency (TF-IDF) [23], Bag-Of-Words (BOWs) [36], neural network word-to-vector embeddings (Word2vec) [18, 36], Part-Of-Speeches (POSs) [5, 11, 36], Context Free Grammers (CFGs) [5, 23, 36], Linguistic Inquiry and Word Count (LIWC) [23], rhetorical relationships among sentences [15, 25], etc.; features can be explicit, i.e., non-latent features such as the frequencies of lexicons, or implicit, i.e., latent features obtained by, for example, WORD2VEC [32, 35]). Recently, news images and source websites have been used in fake news analyses. For example, to investigate news images, Jin et al. [14] defined a set of visual features to predict fake news within a traditional statistical learning framework, and Wang et al. [32] employed a deep neural network (VGG-19) to help extract the latent representation of news images. Baly et al. [1] characterized fake news articles by their source websites, e.g., if they have a Wikipedia page, if their URLs contain digits or domain extensions such as .co, .com, .gov, and their Web traffic information. Nevertheless, few research efforts have focused on the authors who create and write the [true or fake] news, which we investigate in this paper.

On the other hand, fake news detection models have emerged in recent years by studying how fake news propagates on social media (i.e., using social context information). An example can be seen in the work by Vosoughi et al. [31], which revealed that fake news spreads faster, farther, more widely, and is more popular compared to true news. Currently, methods to predict fake news have investigated the profiles of users [7], their social connections [29, 38], and posts. For instance, Guess et al.[7] found that the age of a user is an important indicator of the frequency that he or she engages in fake news activities; the analysis showed that users over 65 shared fake news approximately seven times more often compared to the younger age group. While remarkable progress has been made to achieve early detection of fake news, very little social context information on news propagation may be available; this motivates the development of approaches that can detect fake news by focusing mainly on news content.

A comprehensive survey of the various approaches for handling fake news problem is given in the work by Zhou and Zafarani [37].

Credibility Assessment The credibility of information (including news) is often evaluated by its quality and believability [3]. Research has specifically focused on assessing the credibility of social context information. TweetCred, a real-time system, scores tweet credibility by using a set of hand-crafted features within a semi-supervised learning framework [9]. Out of all selected features, the most important ones include the number of (unique) characters and words in tweets, indicating the significant correlation between the content of tweets and their credibility. In another relevant study, Gupta, et al. [8] found that a majority of the content generated at the time of crisis are from unknown sources (users) and at the same time, rumors are spread, which emphasized the importance of sources on information credibility. Based on a binary classifier, Castillo et al. [3] evaluated information credibility on Twitter using hand-crafted features from users' posting and re-posting behavior, from the text of the posts, and from citations to external sources, which can achieve a precision and recall value between 0.7 and 0.8. Within a similar framework, O'Donovan, et al. [21], discovered that features such as URLs, mentions, retweets, and tweet length were among the best indicators of the credibility using eight diverse Twitter datasets. Morris, et al. [19], found that the name of a Twitter user and the use of standard grammar, influence the credibility of tweets.

In our work, we have adopted features from earlier findings mostly from the context of microblogging sites such as Twitter, to find how closely they relate to news content credibility. We have not found any credibility study focused on number of authors related to the news, authors' collaboration relationships, and authors' past association with fake news articles. The analysis and findings of this paper provide insights to address these issues and improve fake news identification efforts.

We assess credibility from two broad perspectives: (i) Source and (ii) Content which are discussed in details in Sects. 4 and 5. For each category, we identify

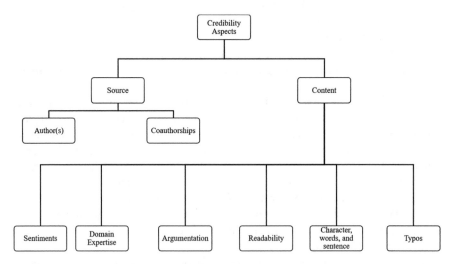

Fig. 1 Hierarchical structure of credibility aspects

Table 1 Data statistics

Data	PolitiFact	BuzzFeed
# Users	23,865	15,257
# News–Users	32,791	22,779
# Users–Users	574,744	634,750
# News stories	240	182
# True news	120	91
# Fake news	120	91

information in fake news that can capture various aspects of credibility, as shown in Fig. 1. Before further elaboration of these categories, we briefly describe the datasets used in our study.

3 Experimental Data

We have used two public datasets for fake news detection, from https://bit.ly/2mHGnBI [28, 29, 36, 38]. These datasets are from Buzzfeed news and Politifact. Besides news content and news labels (i.e., *fake* or *true*), the datasets contain information on the social networks of users involved in spreading the news. Statistics of the two datasets are provided in Table 1.

Out of 422 news articles in both datasets, 16 news articles had the same content/text, and were excluded from our analysis. The datasets were processed using *pandas* [17], and *matplotlib* [12] was used for visualization. In the following sections, we will discuss how different aspects of credibility can be captured from such data and demonstrate our findings on these datasets.

4 Source Credibility

In this section, we present our analysis to derive credibility from the news source, i.e. news URL, number of authors of the news, coauthorship relation to credibility, author(s) affiliations, and history of credibility of authors.

Earlier research has focused on assessing source credibility by looking at the URL associated with any news article [1], including features such as whether a website contains the https prefix, numbers, or .gov, .co, .com domain extensions. In our data, 354 news articles used the http prefix, 15 used the https prefix, and 37 had no URL. Out of the 354 URLs with http prefix, only 154 were fake news. Surprisingly, 14 out of 15 news articles with the https prefix were fake. These observations contradict past studies: having https in a URL does not imply credibility or help differentiate fake news from true news. We also explored whether site names can inform credibility of the news. In the dataset among those which had site names field, we found 87 unique site names with 7 site names common between true and fake news. The total of 112 news of dataset were from these site names with 57 fake and 55 true news. It suggests that distinguishing fake news only based on site names can be extremely difficult as site names of around 27.58 % news could be associated with both fake and true news.

Other studies [3, 8, 9, 19] have shown that specific users information can be good indicators of credibility on Twitter. Hence, we seek generic user information that can capture credibility and help detect fake news. We group such *source* information into information on:

1. *Author(s)* of the news articles, and
2. *Coauthorships*, i.e., author collaborations.

We now discuss each of these subcategories in detail.

4.1 Credibility Signals in Author(s)

If a news article does not provide any information on its authors, its credibility can be questioned. An earlier study found that rumors mostly spread during the times of crises on Twitter and the majority of such rumors are posted by unknown sources/users [8]. However, having the name(s) of the author(s) is insufficient, because fake names or fake profiles can be easily created. Previous work has also looked at whether the news source is Wikipedia or a verified social media account, or contains other attributes to verify its credibility [1].

Thus, credibility assessment methods require multiple steps. First, we simplify credibility assessment to focus on the number of authors in the two types of news, i.e., whether it contains no authors, one author, two authors, or more. We found that the average number of authors is 0.66 for fake news and 1.97 for true news. The median number of authors is 0 for fake news and 2 for true news. Figure 2 provides the distribution of the number of authors for true and fake news.

From Fig. 2, we observe that the number of authors of a news article does have some correlation to its credibility. If an article has more than one author, it is more likely to be credible, and news with no author name is more likely to be fake news. The Pearson correlation coefficient between labels (true/fake) and number of authors is 0.406. It is difficult to draw similar inferences when news articles have only one author. From the figure, we observe that there are almost equal proportions of fake and true news for articles with a single author. A p-value of <0.05 was obtained after running Shapiro-Wilk normality test for the number of authors, which indicates that the distribution is not normal. As the distributions are not normal, we cannot compute the significance in differences in mean values between fake news and true news. Hence, we used Mann-Whitney U test on number of authors on two types of news, i.e., fake and true news. The p-value of <0.001 shows that the median number of authors in these two types of news can capture credibility. In our later analyses, we discuss ways to add past association of the authors with fake news to tackle the case when there is only one author.

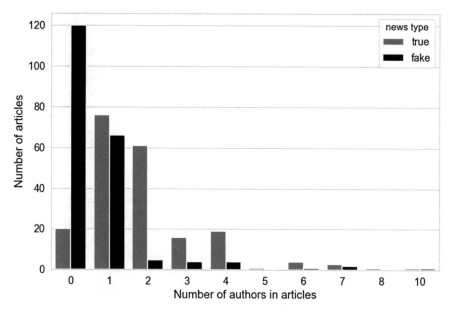

Fig. 2 Frequency of number of authors in fake and true news. If an article has more than one author, it is more likely to be true news

4.2   Credibility Signals in Coauthorships

Earlier studies on coauthorship networks (e.g., Newman [20]) found that (1) a small number of influential individuals exist in such networks and (2) disconnecting such individuals from the network can result in a set of small disjoint networks. Such observations motivate us to explore whether such influential authors exist in the network of news authors, and if they do, can they help assess the credibility of the news, and their coauthors. Hence, we extended our analysis by looking at the network of news authors and classifying them into three groups: authors who are (i) only associated with fake news, (ii) only associated with true news, and (iii) associated with both fake and true news. The objective is to analyze news credibility, given the *position* of the author in this network as well as its *neighbors* (other authors). This approach allows us to understand whether fake news authors collaborate only with other fake news authors, or if they also collaborate with true news authors. We raise similar questions for true news authors and those who publish both.

Among the 237 unique authors in our data, 87 authors were authors of at least two or more news articles. To have sufficient historical data, authors whose names occurred only once (in our data) were excluded from the analysis. For simplicity, we only considered news articles whose authors were in the set of 87 authors. To provide clear insights on coauthorships, we assign these 87 authors to one of three groups:

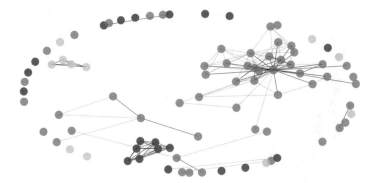

Fig. 3 Authors collaboration network, showing authors that publish only fake news (red), only true news (green), or both (yellow)

1. *True-news authors*, only associated with two or more true news stories;
2. *Fake-news authors*, only associated with two or more fake news stories;
3. *Fake+True authors*, who have published both fake news and true news.

For these groups, the coauthorship network among authors is shown in Fig. 3, where green nodes represent *True-news authors*, red nodes represent *Fake-news authors*, and yellow nodes represent *Fake+True authors*. Dashed lines connect authors that have collaborated only once, whereas solid lines connect authors who have collaborated more than once. We notice that only 12.7% of 87 authors are involved in both fake and true news, whereas the majority were either exclusively involved in fake news or true news. We also observe that fake news authors are often either the only author (of the fake news article), or are more likely to collaborate with other *Fake-news authors* (rather than with *True-news authors* or *Fake+True authors*). We had similar observations for *True-news authors* (green nodes in Fig. 3) and *Fake+ True authors* (yellow nodes in Fig. 3).

To further investigate these observations, for each author in the coauthorship graph, we compute the number of coauthors (i.e., graph neighbors) who only post true news, only post fake news, and those who post both. Using these three numbers, we can represent any author as a 3D point and plot all authors in 3D space, as shown in Fig. 4. We observe that the credibility of authors who have had multiple coauthorhips are easily distinguishable, as they often collaborate with the same type of authors. Hence, knowing the author's credibility, we can infer the credibility of coauthors. For authors with no neighbors (i.e., coauthors), they are indistinguishable.

In sum, *homophily* exist in authorship [34], where authors who write only true news are less likely to collaborate with authors who write fake news. These observations also indicate that if there are groups of authors associated with some news, by knowing credibility of any author, we may be able to infer the credibility of the news and its other authors. But, *how can we determine an author's credibility?* Two observations help us address this question:

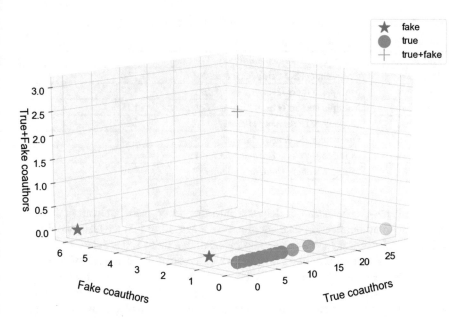

Fig. 4 3D plot of authors. An author is represented in terms of three values: the number of coauthors of him or her who only post true news, fake news, or both

I. **Affiliations provide information on credibility.** Some author names are associated with known organizations such as ABC news, Associated Press, Politico, and CNN.

Hence, affiliations of authors with well-recognized organizations may indicate that a news article is not fake. While we did not explicitly consider the author relationships with the organizations, these findings support the hypothesis that working for a credible organization allows one to infer author's credibility. Similarly, we found some unrealistic author names such as Fed up, Ny evening, About the Potatriot, and About Stryker associated with fake news. This finding corroborates the earlier observation by Gupta, et al. [8], where rumors on Twitter were shown to have been spread by unknown sources/users.

II. **Historical record provides information on credibility.** The 87 authors that were selected were related to 172 news articles. In Table 2, we looked at how different types of authors relate to the two types of news. We observe that around 28% of the news articles have authors who post both fake and true news. However it is unclear how this information can help infer credibility of the authors or the news. To tackle this issue, we looked into the history of authors' credibility, i.e., their past associations with true/fake news articles, in order to explore whether these can capture the credibility of other articles authored later by the same authors.

Table 2 Distribution of count of news articles based on coauthors

News type	Number of authors (author type(s))	Number of articles
True news	One author (*True-news authors*)	37
True news	One author (*Fake+True news authors*)	18
True news	Multiple authors (*True-news authors*)	59
True news	Multiple authors (*Fake+True news authors*)	6
Fake news	One author (*Fake news authors*)	23
Fake news	One author (*Both news authors*)	18
Fake news	Multiple authors (*Fake news authors*)	5
Fake news	Multiple authors (*Both news authors*)	6

Not all news articles in our data had information on their publication date. As articles with no publication date could not help with historical analysis, we filtered them, reducing our data to 289 news articles. Among these 289 articles, we focused on authors of at least two news articles, which resulted in 69 authors of 163 news articles. For each author, their published news articles were sorted chronologically and we analyzed whether they contradicted their past behavior anytime in the future. We only found 11 authors that contradicted, i.e., either they had fake news in the past and were associated with true news in future, or they posted true news in the past but were associated with fake news in the future. However, the majority of authors (84%) showed consistent behavior. Thus, past information on authors' credibility provides insights on the credibility of other articles authored by them.

5 Content Credibility

Next, we discuss credibility based on the content of the news. Our goal is to explore how various characteristics of a news article content (which includes the text from the title and the body of an article) can help assess its credibility. These characteristics are compared for fake and true news articles. Previous research on credibility on Twitter [3, 9, 19, 21] has shown that there exist various indicators of credibility within content. Here, we search for such indications of credibility in the following: (1) sentiments expressed, (2) domain expertise in the article, (3) arguments used, (4) text readability, (4) characters, words, and sentences used, and (5) typos.

5.1 Credibility Signals in Sentiments

Castillo et al. [3] identified connections between sentiments expressed and credibility, whereas O'Donovan, et al. [21], found that positive sentiments may not indicate credibility in tweets. Such studies encouraged us to study the relationship between credibility and sentiments in news, answering questions such as:

- Are sentiments expressed in fake news different from those in true news?
- Is fake news written with negative, neutral, or positive sentiments?
- Can sentiments help infer credibility?

For sentiment analysis, we compute sentiment intensity (a numeric value) for each sentence in the news articles using VADER [13] sentiment analyzer available in NLTK (natural language toolkit) [2]. Using the standard threshold [13], three labels are assigned to each sentence: positive, negative, or neutral. We represent each news article with the three fractions of negative, positive, and neutral sentences in the article, e.g., $\frac{\text{number of positive sentences}}{\text{total number of sentences}}$. Some statistics on such proportions of each type of sentiment in fake and true news are in Table 3.

The mean and median values in Table 3 show that (1) proportion of neutral sentiments is slightly higher in true news compared to fake news and (2) negative sentiments proportions are higher in fake news compared to true news. However, it is still difficult to infer whether sentiments are good indicators of credibility. For further analysis, we explored whether sequence of expressed sentiments in articles differ, i.e. is fake news more likely to have sequences of sentences with positive sentiments followed by other sentences with positive sentiments? There are 9 (3×3) possible types of sequences that one can get with positive, negative, and neutral sentences. We label each sentence pair as one of these types. The mean and median proportion for each one of these nine types in both fake and true news are provided in Table 4.

Table 4 shows that positive sentences that are immediately followed by neutral sentences are more in true news compared to fake news. In fake news, negative sentences followed by other sentiments occur more often than in true news. Overall,

Table 3 Sentiment proportions in news

News type	Positive		Neutral		Negative	
	Mean	Median	Mean	Median	Mean	Median
Fake news	0.34	0.32	0.28	0.27	0.38	0.38
True news	0.33	0.33	0.34	0.31	0.34	0.33

Table 4 Statistics on sequences of sentences with different sentiments

	Mean		Median	
Sentence sequence	Fake news	True news	Fake news	True news
Positive sentence followed by a positive sentence	0.14	0.13	0.1	0.11
Positive sentence followed by negative sentence	0.10	0.09	0.10	0.09
Positive sentence followed by neutral sentence	0.09	0.11	0.08	0.10
Negative sentence followed by positive sentence	0.11	0.09	0.10	0.09
Negative sentence followed by negative sentence	0.16	0.14	0.13	0.11
Negative sentence followed by neutral sentence	0.11	0.12	0.10	0.11
Neutral sentence followed by positive sentence	0.08	0.09	0.07	0.10
Neutral sentence followed by negative sentence	0.09	0.08	0.09	0.07
Neutral sentence followed by neutral sentence	0.09	0.13	0.06	0.10

the uniformity of values in Table 4 on sentiment sequences in news articles suggests that sentiment sequences may be weak indicators of credibility. In contrast to earlier findings [3, 21], our results show that relying on sentiments alone may provide only a weak indication of credibility.

5.2   Credibility Signals in Domain Expertise

Research has shown that the presence of signal words and expert sources enhance credibility [33]. As our data was collected during the U.S. 2016 presidential election, we looked into the use of words from NCSL (National Conference of State Legislatures), which included 150 words from https://bit.ly/1iMTzXa. We studied whether there exist differences between fake news and true news in terms of usage frequencies of these words. We found that the average number of words from the NCSL word list were 4.37 in fake news and 7.46 in true news. The medians for the number of words were 3 for fake news, and 4 for true news. The Shapiro-Wilk test on the number of NCSL words on both fake and true news had a p-value of < 0.05, showing the sample is not normally distributed on number of words for both types of news. With a small difference of one word, it is difficult to argue the importance of domain words/phrases, so, we further looked into distinct words that are present in one type of news and not in the other, shown in Table 5.

Later in our experiments, we will show how the occurrence of words shown in Table 5, in addition to other information, allows one to detect fake news.

5.3   Credibility Signals in Argumentation

To build strong arguments in a news article, one can rely on providing data and references.

Frequent occurrences of numbers or digits may indicate that a news article is well-researched, containing verifiable data; similarly, the occurrence of hyperlinks and URLs may indicate citations suggesting that an article is supported by external sources. The connections between URLs and credibility have been studied earlier on tweets [3, 9]. Similarly, the findings from Koetsenruijter [16] suggests that the

Table 5 List of NCSL words in fake and true news

Words in fake news, but not in true news	`petition, legislator, impeachment, adhere`
Words in true news, but not in fake news	`fiscal, calendar, precedent, bipartisan, convene, interim, caucus, nonpartisan, statute, decorum, veto, repeal, constituent, chamber`

Table 6 Distribution of
digits in fake and true news

Mean		Median	
Fake news	True news	Fake news	True news
490.82	739.33	424	461

presence of numbers in an article conveys credibility. Hence, we studied whether there is a difference between fake news and true news in terms of numbers of URLs.

We found that only 18 news articles contained URLs, of which 7 were fake news and 11 were true news. With presence of URL in small proportion of dataset, it is difficult to assess credibility strength based on this feature. Table 6 shows the distribution of number of digits used in fake and true news. Our findings suggest that there are differences between fake news and true news based on the use of numbers in the news content, and that it is likely that true news is supported with facts that include numbers.

The Shapiro-Wilk test gave a p-value of <0.05, showing that the sample is not normally distributed on number of words for both type of news. The Mann-Whitney U test shows that there is a difference in medians with a p-value of 0.011, i.e., the greater occurrence of digits in news articles indicates credibility.

5.4 Credibility Signals in Readability

The study by Horne et al. [11] suggested readability as an important feature to distinguish fake news from true news. To compare readability differences between fake and true news, we used the Flesch-Kincaid reading-ease test on the text of the news. The mean readability scores were found to be 67.32 for fake news and 65.30 for true news. Similarly, the median scores are 68.33 and 65.38 for fake news and true news, respectively. Contrary to our expectation, fake news readability was higher than that of true news. This raises a series of other interesting questions such as:

- Is fake news more readable?
- Is ease of reading why users engage more with fake news than true news?

Further analysis of the news content may reveal insights on such questions. The Shapiro-Wilk test, with p-value <0.05 indicates that the sample is not normally distributed. The Mann-Whitney U test with a p-value of 0.02 shows differences in medians and that poorer readability may indicate credibility.

5.5 Credibility Signals in Characters, Words, and Sentences

In TweetCred system [9], the number of characters and number of words were among the important features to evaluate credibility of tweets. Earlier work has also shown that tweet length is one of the indicators for credibility [9, 21]. Hence,

Table 7 Distribution of words in title and text

News type	# Words in title		# Words in text		# Sentences		# Words / # Sentences	
	Mean	Median	Mean	Median	Mean	Median	Mean	Median
Fake news	14.16	14.0	490.82	424.0	19.44	16.0	26.80	25.32
True news	12.09	11.0	739.33	461.0	26.59	17.0	27.69	27.50

Table 8 Distribution of characters and words

News type	# Characters		# Characters / # Words		# Special characters	
	Mean	Median	Mean	Median	Mean	Median
Fake news	2052.76	1803.0	4.23	4.20	6.79	4.0
True news	3144.28	1899.0	4.27	4.27	11.04	6.0

the news content length may also be an indicator for the credibility. The mean and median of words in title and text of the news are shown in Table 7 along with the number of words per sentences. Table 7 shows that fake news text is shorter in terms of number of words compared to true news. Similarly, the length of true news articles and their number of sentences were found to be higher on average compared to fake news.

In similar research, it has been shown that presence of special characters, e.g., colon [9], exclamation mark, and question mark [3], can help assess credibility. To check if fake news contains more special characters compared to true news, we select !, #, $, %, *, +, -, ?, @, | as special characters and count their occurrences in each type of news. Table 8 provides the mean and median of number of characters and special characters in news text, indicating that special characters are more often observed in true news.

5.6 Credibility Signals in Typos

As suggested by Morris et al. [19], the use of standard grammar and spellings enhances credibility. So, we checked if there are significant differences in typographical errors between fake and true news. To find if there are more typos in fake news compared to true news, we used the words from the NLTK corpus, containing 235,892 English words. For each news content, we counted the number of words with typos and normalized it by the total number of words in the content. Words from both content and NLTK corpus were lower-cased before checking for typos. Contrary to our expectation, we found on average 0.19 and 0.22 typos in fake news and true news, respectively. The median number of typos were 0.20 for fake news and 0.21 for true news. The Shapiro-Wilk test indicated that the sample is not normally distributed, and using Mann-Whitney U test, we obtained a p-value <0.001, showing that typos may indicate credibility.

6 Results

Based on our discussion on credibility aspects in Sects. 4 and 5, we built different fake news prediction models to predict fake news. From the attributes discussed, we obtained 26 features in the following categories:

- Number of authors in the news;
- Sentiments (counts of positive, neutral, and negative sentiments in text, and sequence of sentiments);
- Number of NCSL words that are only present in fake news;
- Number of NCSL words that are only present in true news;
- Flesch-Kincaid reading-ease score;
- Number of words in the title;
- Number of characters, special characters, words, sentences, digits, and typos;
- Words per sentences;
- Characters per words; and
- Past history of the author.

Using the above features, we trained a fake news prediction model with seven different classifiers (to account for learning bias) using scikit-learn package [22] and ten-fold cross validation. The F_1-scores for these classifiers are in Table 9, where we found that the logistic regression and linear Support Vector Machines performed well among different classifiers. The best classification was achieved by logistic regression, with an 0.80 average F_1-macro score.

Comparing Source-Credibility and Content-Credibility We studied the importance of assessing each type of credibility (source and content) by predicting fake news independently using each category of features. For source-credibility, we only considered three features: number of authors, as well as the numbers of past fake and true news stories authored by them in the past. Surprisingly, with these three

Table 9 Average F_1 scores for all features (highest values are shown in bold font)

Classifier	F_1-micro	F_1-macro	F_1-weighted
SVM (RBF kernel)	0.74	0.74	0.74
Linear SVM	0.79	0.79	0.79
Logistic regression	**0.80**	**0.80**	**0.80**
Random forest	0.76	0.76	0.76
AdaBoost	0.74	0.74	0.74
Naive bayes	0.69	0.69	0.69
Gradient boosting decision tree	0.77	0.77	0.77

Table 10 Average F_1 scores obtained by source-credibility features (highest values are shown in bold font)

Classifier	F_1-micro	F_1-macro	F_1-weighted
SVM (RBF kernel)	0.75	0.75	0.75
Linear SVM	0.75	0.75	0.75
Logistic regression	0.75	0.74	0.74
Random forest	0.76	0.76	0.76
AdaBoost	**0.77**	**0.77**	**0.77**
Naive bayes	0.75	0.75	0.75
Gradient boosting decision tree	**0.77**	0.76	0.76

Table 11 Average F_1 scores obtained by content-credibility features (highest values are shown in bold font)

Classifier	F_1-micro	F_1-macro	F_1-weighted
SVM (RBF kernel)	0.64	0.63	0.63
Linear SVM	**0.68**	**0.68**	**0.68**
Logistic regression	0.67	0.67	0.67
Random forest	0.63	0.63	0.63
AdaBoost	0.60	0.60	0.60
Naive bayes	0.58	0.57	0.57
Gradient boosting decision tree	0.65	0.65	0.65

features, we find that the classification performance does not degrade much, as shown in Table 10. However, the best classifier was then AdaBoost, which indicates that the classifier performance is feature-dependent. Similarly, when using the 23 content-credibility features, the best F_1-score achieved was 0.68, less than when using only source-credibility features, which achieved 0.77. The results are shown in Table 11.

By comparing the performance of source-credibility and content-credibility features, we find that assessing source credibility plays a stronger role in detecting fake news. Adding content-credibility features with source-credibility features can further improve fake news detection.

Feature Importance Analysis We also identified the most important features that can capture credibility in news.

While there can be various combinations of features to search for the optimal features, we combined both the feature selection and a hand-tailored approach (testing with trial and error), which led to 13 features with the best F_1 score: number of authors in the news and past history of authors, presence of domain words, readability, number of words, characters, special characters, and typographical errors. Table 12 shows that all classifiers performed best with these selected features, even better than using the original 26 features. Also, features that were found to be of least importance were sentiments and count of digits in the text.

Table 12 Average F_1 score with only 13 features (highest values are shown in bold font)

Classifier	F_1-micro	F_1-macro	F_1-weighted
SVM (RBF kernel)	0.77	0.77	0.77
Linear SVM	**0.80**	**0.80**	**0.80**
Logistic regression	**0.80**	**0.80**	**0.80**
Random forest	0.77	0.77	0.77
AdaBoost	0.77	0.77	0.77
Naive bayes	0.75	0.75	0.75
Gradient boosting decision tree	0.77	0.77	0.77

Table 13 Average F_1 score using 13 features on separated datasets (highest values are shown in bold font)

Classifier	F_1-macro (Politifact)	F_1-macro (Buzzfeed)
SVM (RBF kernel)	0.79	0.75
Linear SVM	**0.82**	**0.77**
Logistic regression	**0.82**	0.76
Random forest	0.77	0.72
AdaBoost	0.78	0.67
Naive bayes	0.79	0.69
Gradient boosting decision tree	0.80	0.74

Comparing Tables 9 and 12, we can observe that using these 13 selected features, all the classifiers perform better than using the original 26 features. While our results did not outperform all other models as discussed in [36], where best model had a 0.892 F_1 score, our model used comparatively fewer and new features compared to the models discussed in the work.

With only 3 source-credibility features on Politifact data, the best classifier achieved an average F_1-macro score of 0.83 and with only content-credibility features, best score was 0.66 (see Table 13). This observation shows that content-credibility has very little to add to the performance of fake news prediction in the data. Similarly, for Buzzfeed news data, the best classifier was able to obtain an average F_1-macro score of 0.76 with only source-credibility features, whereas with content-credibility features it obtained 0.66. Thus, adding content-credibility features only slightly improved the performance. Our content-credibility features are comparatively fewer than earlier studies, so we emphasize our findings with source-credibility features, which we did not find in earlier research.

7 Conclusion

We have analyzed credibility of news, emphasizing features related to source and content of the articles. Our results based on source of the news (Sect. 4) show that number of authors of the news is a strong indicator of credibility. We found that when the news article has no authors, it is more likely to be fake news. Our findings on collaboration of authors suggests that authors who are engaged in true credible news are less likely to collaborate with authors who are associated with fake news. This indicates that for a news article with multiple authors, by knowing the credibility of one author, we can infer the credibility of the news as well as other coauthors. Furthermore, we found that authors' affiliations with well-recognized organizations can be a signal for credibility. The results also suggest that credibility history of authors can provide insights on credibility of other articles from the same author.

Similarly, we investigated credibility based on various content-related aspects of the news (Sect. 5). The results show that sentiments expressed in news articles are weak indicators of credibility. We observed that the use of numbers in true news articles occurred more often than in fake news, perhaps because true news is supported with facts that include numbers. Comparing the number of words and sentences in true news and fake news showed that on average, true news had more words and sentences than fake news. Surprisingly, we observed more typos in true news than in fake news. Our analyses also showed that domain expertise on topics discussed in news can enhance fake news detection.

After analysis of individual features, we used our findings to build predictive models to detect fake news. The F_1-score of 0.80 obtained by predictive models built with source-credibility features show that with a small number of features, one can still can detect fake news reasonably well. Using fewer features can lead to less complex models. Hence, our simple approach provides a straightforward fake news detection framework with a few features that can quickly detect fake news.

Stronger conclusions require further research on additional machine learning features, other predictive models, and datasets. We have not yet explored word-based sentiments in our analysis, where one can consider negated positive words, or number of negative and positive words in sentences. Another avenue to explore is to study sentiments in paragraphs, which may show less variation compared to our results. Furthermore, the news content can include images (and other media), as well as the number of user interactions, which may provide more insights on the differences between fake and true news.

References

1. Baly, R., Karadzhov, G., Alexandrov, D., Glass, J., Nakov, P.: Predicting factuality of reporting and bias of news media sources. arXiv preprint arXiv:1810.01765 (2018)
2. Bird, S., Klein, E., Loper, E.: Natural Language Processing with Python: Analyzing Text With the Natural Language Toolkit. O'Reilly Media, Inc, Beijing (2009)
3. Castillo, C., Mendoza, M., Poblete, B.: Information credibility on twitter. In: Proceedings of the 20th International Conference on World Wide Web, pp. 675–684. ACM (2011)
4. Ciampaglia, G.L., Shiralkar, P., Rocha, L.M., Bollen, J., Menczer, F., Flammini, A.: Computational fact checking from knowledge networks. PLoS One **10**(6), e0128193 (2015)

5. Feng, S., Banerjee, R., Choi, Y.: Syntactic stylometry for deception detection. In: Proceedings of the 50th Annual Meeting of the Association for Computational Linguistics: Short Papers, vol. 2, pp. 171–175. Association for Computational Linguistics (2012)
6. Fletcher, R., Nielsen, R.K.: Are people incidentally exposed to news on social media? a comparative analysis. New Media Soc. **20**(7), 2450–2468 (2018)
7. Guess, A., Nagler, J., Tucker, J.: Less than you think: prevalence and predictors of fake news dissemination on facebook. Sci. Adv. **5**(1), eaau4586 (2019)
8. Gupta, A., Kumaraguru, P.: Twitter explodes with activity in Mumbai blasts! a lifeline or an unmonitored daemon in the lurking? Technical report (2012)
9. Gupta, A., Kumaraguru, P., Castillo, C., Meier, P.: Tweetcred: real-time credibility assessment of content on twitter. In: International Conference on Social Informatics, pp. 228–243. Springer (2014)
10. Hassan, N., Arslan, F., Li, C., Tremayne, M.: Toward automated fact-checking: detecting check-worthy factual claims by claimbuster. In: Proceedings of the 23rd ACM SIGKDD International Conference on Knowledge Discovery and Data Mining, pp. 1803–1812. ACM (2017)
11. Horne, B.D., Adali, S.: This just in: fake news packs a lot in title, uses simpler, repetitive content in text body, more similar to satire than real news. In: Eleventh International AAAI Conference on Web and Social Media (2017)
12. Hunter, J.D.: Matplotlib: A 2d graphics environment. Comput. Sci. Eng. **9**(3), 90 (2007)
13. Hutto, C.J., Gilbert, E.: Vader: a parsimonious rule-based model for sentiment analysis of social media text. In: Eighth International AAAI Conference on Weblogs and Social Media (2014)
14. Jin, Z., Cao, J., Zhang, Y., Zhou, J., Tian, Q.: Novel visual and statistical image features for microblogs news verification. IEEE Trans. Multimedia **19**(3), 598–608 (2017)
15. Karimi, H., Tang, J.: Learning hierarchical discourse-level structure for fake news detection. arXiv preprint arXiv:1903.07389 (2019)
16. Koetsenruijter, A.W.M.: Using numbers in news increases story credibility. Newspaper Res. J. **32**(2), 74–82 (2011)
17. McKinney W.: Data Structures for Statistical Computing in Python, Proceedings of the 9th Python in Science Conference, 51–56 (2010)
18. Mikolov, T., Chen, K., Corrado, G., Dean, J.: Efficient estimation of word representations in vector space. arXiv preprint arXiv:1301.3781 (2013)
19. Morris, M.R., Counts, S., Roseway, A., Hoff, A., Schwarz, J.: Tweeting is believing? understanding microblog credibility perceptions. In: Proceedings of the ACM 2012 Conference on Computer Supported Cooperative Work, pp. 441–450. ACM (2012)
20. Newman, M.E.: Coauthorship networks and patterns of scientific collaboration. Proc. Natl. Acad. Sci. **101**(suppl 1), 5200–5205 (2004)
21. O'Donovan, J., Kang, B., Meyer, G., Höllerer, T., Adalii, S.: Credibility in context: an analysis of feature distributions in twitter. In: 2012 International Conference on Privacy, Security, Risk and Trust and 2012 International Conference on Social Computing, pp. 293–301. IEEE (2012)
22. Pedregosa, F., Varoquaux, G., Gramfort, A., Michel, V., Thirion, B., Grisel, O., Blondel, M., Prettenhofer, P., Weiss, R., Dubourg, V., Vanderplas, J., Passos, A., Cournapeau, D., Brucher, M., Perrot, M., Duchesnay, E.: Scikit-learn: machine learning in Python. J. Mach. Learn. Res. **12**, 2825–2830 (2011)
23. Pérez-Rosas, V., Kleinberg, B., Lefevre, A., Mihalcea, R.: Automatic detection of fake news. arXiv preprint arXiv:1708.07104 (2017)
24. Potthast, M., Kiesel, J., Reinartz, K., Bevendorff, J., Stein, B.: A stylometric inquiry into hyperpartisan and fake news. arXiv preprint arXiv:1702.05638 (2017)
25. Rubin, V.L., Lukoianova, T.: Truth and deception at the rhetorical structure level. J. Assoc. Inf. Sci. Technol. **66**(5), 905–917 (2015)
26. Shi, B., Weninger, T.: Discriminative predicate path mining for fact checking in knowledge graphs. Knowl. Based Syst. **104**, 123–133 (2016)

27. Shu, K., Mahudeswaran, D., Wang, S., Lee, D., Liu, H.: Fakenewsnet: a data repository with news content, social context and dynamic information for studying fake news on social media. arXiv preprint arXiv:1809.01286 (2018)
28. Shu, K., Sliva, A., Wang, S., Tang, J., Liu, H.: Fake news detection on social media: a data mining perspective. ACM SIGKDD Explor. Newsletter **19**(1), 22–36 (2017)
29. Shu, K., Wang, S., Liu, H.: Beyond news contents: the role of social context for fake news detection. In: Proceedings of the Twelfth ACM International Conference on Web Search and Data Mining, pp. 312–320. ACM (2019)
30. Silverman, C.: This analysis shows how viral fake election news stories outperformed real news on facebook (2016). https://www.buzzfeednews.com/article/craigsilverman/viral-fake-election-news-outperformed-real-news-on-facebook#.vtQpz9DKd
31. Vosoughi, S., Roy, D., Aral, S.: The spread of true and false news online. Science **359**(6380), 1146–1151 (2018)
32. Wang, Y., Ma, F., Jin, Z., Yuan, Y., Xun, G., Jha, K., Su, L., Gao, J.: Eann: event adversarial neural networks for multi-modal fake news detection. In: Proceedings of the 24th ACM SIGKDD International Conference on Knowledge Discovery & Data Mining, pp. 849–857. ACM (2018)
33. Wogalter, M.S., Kalsher, M.J., Rashid, R.: Effect of signal word and source attribution on judgments of warning credibility and compliance likelihood. Int. J. Ind. Ergon. **24**(2), 185–192 (1999)
34. Zafarani, R., Abbasi, M.A., Liu, H.: Social media mining: an introduction. Cambridge University Press, New York (2014)
35. Zhang, J., Cui, L., Fu, Y., Gouza, F.B.: Fake news detection with deep diffusive network model. arXiv preprint arXiv:1805.08751 (2018)
36. Zhou, X., Jain, A., Phoha, V.V., Zafarani, R.: Fake news early detection: a theory-driven model. arXiv preprint arXiv:1904.11679 (2019)
37. Zhou, X., Zafarani, R.: Fake news: a survey of research, detection methods, and opportunities. arXiv preprint arXiv:1812.00315 (2018)
38. Zhou, X., Zafarani, R.: Network-based fake news detection: a pattern-driven approach. arXiv preprint arXiv:1906.04210 (2019)

Standing on the Shoulders of Guardians: Novel Methodologies to Combat Fake News

Nguyen Vo and Kyumin Lee

Abstract Fake news and misinformation are one of the most pressing issues of modern society. In fighting against fake news, many fact-checking systems such as human-based fact-checking sites (e.g., snopes.com and politifact.com) and automatic detection systems have been developed in recent years. However, online users still keep sharing fake news even when it has been debunked. It means that early fake news detection may be insufficient and we need complementary approaches to mitigate the spread of misinformation. In this chapter, we introduce novel methods to intervene the spread of fake news and misinformation. In particular, we (1) leverage online users named *guardians*, who cite fact-checking sites as credible evidences to fact-check information in public discourse, (2) propose two novel frameworks – the first one is a recommender system to personalize fact-checking articles[1] and the second one is a text generation framework[2] to generate responses with fact-checking intention. Both frameworks are designed to increase the guardians' engagement in fact-checking activities. Experimental results showed that our recommender system improves competitive baselines significantly by 10~20%, and the text generation framework is able to generate relevant responses and outperforms state-of-the-art models by achieving up to 30% improvement. Our qualitative study also confirms that the superiority of our generated responses compared with responses generated from the existing models.

Keywords Fact-checking · Fake news · Fact-checkers · Recommendation · Text generation

---

[1] https://github.com/nguyenvo09/CombatingFakeNews

[2] https://github.com/nguyenvo09/LearningFromFactCheckers

N. Vo · K. Lee (✉)

Department of Computer Science, Worcester Polytechnic Institute, Worcester, MA, USA

e-mail: nkvo@wpi.edu; kmlee@wpi.edu

K. Shu et al. (eds.), *Disinformation, Misinformation, and Fake News in Social Media*, Lecture Notes in Social Networks,
https://doi.org/10.1007/978-3-030-42699-6_10

1   Introduction

Our media landscape has been flooded by a large volume of falsified information, overstated statements, false claims, fauxtography and fake videos[3] perhaps due to the popularity, impact and rapid information dissemination of online social networks. The dramatic increase in the volume of misinformation posed severe threats to our society, degraded trustworthiness of cyberspace, and influenced the physical world. For example, $139 billion was wiped out when the Associated Press (AP)'s hacked Twitter account posted fake news regarding White House explosion with Barack Obama's injury. Owing to the detrimental impact on modern society, a large body of research work and efforts have been focused on detecting fake news and building online fact-check systems in order to debunk fake news in its early stage of dissemination.

However, falsified news is still disseminated like wild fire [19, 33] despite the rise of fact-checking sites worldwide in the last half decade [11]. One possible explanation for the aforementioned phenomenon is that verifying the correctness of information may not be a common practice of the majority of people[4] since it takes time to search and read lengthy fact-checking articles. Furthermore, recent work showed that individuals tend to selectively consume news that have ideologies similar to what they believe while disregarding contradicting arguments [7, 21]. These reasons and problems indicate that using only fact-checking systems to debunk fake information is insufficient, and complementary approaches are necessary to combat fake news.

Therefore, in this chapter, we focus on online users named *guardians*, who directly engage with other users in public dialogues and convey verified information to them. Figure 1 shows a real-life conversation between two online users. The user @TheRightMelissa, called *original poster*, posts fake news about the wall between Guatemala and Mexico. After few minutes, the user @EmmaDaly refutes the misinformation by replying to the original poster and provides a fact-checking article as a supporting evidence. We call such a reply a *Direct Fact-checking tweet* (D-tweet) and the user who posts the D-tweet is called a *D-guardian*. Additionally, we notice that the D-tweet is retweeted eleven times. We call users who retweet the D-tweet secondary guardians (S-guardians) and their retweets are called secondary fact-checking tweets (S-tweets). Both *D-guardians* and *S-guardians* are called *guardians*, and both D-tweets and S-tweets are named fact-checking tweets.

In Sect. 2.1, we will show that guardians often quickly fact-checked original tweets within a day after being posted and their D-tweets could reach hundreds of millions of followers. Additionally, the likelihood to delete shares of fake news

---

[3]https://cnnmon.ie/2AWCCix

[4]http://go.zignallabs.com/Q1-2017-fake-news-report

Melissa A.
@TheRightMelissa

Mexico's own southern border wall with Guatemala.
Only MSM brainwashed fools think it's racist to want to
protect your country #NoBanNoWall bit.ly/2ZtWuAB

January 25, 2017 4:09:37 PM

1.8K Retweets   **2.1K** Likes

EmmaDaly @EmmaDaly · January 25, 2017 4:13:47 PM
Replying to @TheRightMelissa

@TheRightMelissa @sweetatertot2 No, that's part of Israel's West
Bank barrier #NoBanNoWall t.co/GY9UdpAWq8

11 Retweets   **53** Likes

Fig. 1 A real-life fact-checking activity where the D-guardian @EmmaDaly refutes misinformation in the original tweet about the wall between Guatemala and Mexico after the original tweet was posted in few minutes

increased by 4 times when there existed a fact-checking URL in users' comments [8].

Due to the guardians' activeness and high impact on dissemination of fact-checked content, our goal is to further support them in fact-checking activities toward complementing existing fact-checking systems and combating fake news. In particular, we propose (1) a novel fact-checking URLs recommendation to recommend new and interesting fact-checking articles to guardians and (2) build a text generation framework to generate responses with fact-checking intention when original tweets are given. The fact-checking intention means either confirming or refuting content of an original tweet by providing credible evidences. Regarding two goals, this chapter shall describe these frameworks as novel methods to combat fake news.

2   Fact-Checking Article Recommendation System

In this section, we investigate who guardians are, their activeness in fact-checking activities and their impact in disseminating fact-checked contents. Based on guardians' posted fact-checking articles, we build our recommender system to personalize these articles as a way to improve guardians' engagement in fact-checking activities.

2.1   Data Collection

We employed the Hoaxy system [26] to collect a large number of D-tweets and S-tweets. In particular, we collected 231,377 unique *fact-checking tweets* from six well-known fact-checking websites – *Snopes.com*, *Politifact.com*, *FactCheck.org*, *OpenSecrets.org*, *TruthOrfiction.com* and *Hoax-slayer.net* – via the APIs provided by the Hoaxy system which internally used Twitter streaming API. The collected data consisted of 161,981 D-tweets and 69,396 S-tweets (58,821 retweets of D-tweets and 10,575 quotes of D-tweets) generated from May 16, 2016 to July 7, 2017 (~1 year and 2 month).

We removed tweets containing only base URLs (e.g., snopes.com or politi-fact.com) or URLs simply pointing to the background information of the websites because the tweets containing these URLs may not contain fact-checking information. After filtering, we had 225,068 fact-checking tweets consisting of 157,482 D-tweets and 67,586 S-tweets posted by 70,900 D-guardians and 45,406 S-guardians. 7,167 users played both roles of D-guardians and S-guardians. The number of unique fact-checking URLs was 7,295. In addition, we collected each guardian's recent 200 tweets. Table 1 shows the statistics of the collected dataset.

2.2   Identities of Guardians and their Activeness

As we have shown in the previous section, there were only 7,167 users (7%) who behaved as both D-guardians and S-guardians, which indicates that guardians usually focused on either fact-checking claims in conversations (i.e., being D-guardians) or simply sharing credible information (i.e., being S-guardians). Since D-guardians and S-guardians played different roles, we seek to understand which group is more enthusiastic about its role. We created two lists – a list of the number of D-tweets posted by each D-guardian and a list of the number of S-tweets posted by each S-guardian –, excluding D&S guardians who performed both roles. Then, by conducting One-sided MannWhitney U-test, we found that D-guardians were significantly more enthusiastic about their role than S-guardians (p-value<10^{-6}). We also found that even the D&S guardians posted relatively larger number of D-tweets than S-tweets according to Wilcoxon one-sided test (p-value<10^{-6}).

The majority of guardians (85.3%) posted only 1~2 fact-checking tweets. However, there were active guardians, each of whom posted over 200 fact-checking tweets. Tables 2 and 3 show the top 15 most active D-guardians and S-guardians and the number of their D-tweets and S-tweets. Red-colored *Jkj193741* and *upayr*

Table 1 Statistics of our dataset

|D-tweets|	|S-tweets|	|D-guardians|	|S-guardians|	|D&S guardians|
157,482	67,586	70,900	45,406	7,167

Table 2 Top 15 most active D-guardians and associated number of D-tweets

D-guardians and their |D-tweets|		
RandoRodeo (450)	stuartbirdman (318)	upayr (214)
pjr_cunningham (430)	ilpiese (297)	JohnOrJane (213)
TXDemocrat (384)	BreastsR4babies (255)	GreenPeaches2 (199)
Jkj193741 (355)	rankled2 (230)	spencerthayer (195)
BookRageStuff (325)	___lor__ (221)	SaintHeartwing (174)

Table 3 Top 15 most active S-guardians, and associated number of S-tweets

S-guaridans and their |S-tweets|		
Jkj193741 (294)	MrDane1982 (49)	LeChatNoire4 (35)
MudNHoney (229)	pinch0salt (46)	bjcrochet (34)
_sirtainly (75)	ActualFlatticus (42)	upayr (33)
Paul197 (66)	BeltwayPanda (36)	58isthenew40 (33)
Endoracrat (49)	EJLandwehr (36)	slasher48 (31)

Table 4 Top 15 verified guardians, and corresponding D-tweet and S-tweet count

Verified guardians and (|D-tweets| vs. |S-tweets|)		
fawfulfan (103-1)	tomcoates (37-0)	KimLaCapria (27-3)
OpenSecretsDC (37-30)	**aravosis (29-8)**	PattyArquette (29-0)
PolitiFact (41-17)	**TalibKweli (27-8)**	NickFalacci (28-0)
RobertMaguire_ (46-7)	rolandscahill (31-0)	AaronJFentress (28-0)
jackschofield (42-1)	MichaelKors (30-0)	ParkerMolloy (26-1)

guardians were especially active in joining online conversations and spreading fact-checked information.

Next, we examined whether guardians have *verified* Twitter accounts or are highly visible users, who have at least 5,000 followers. The verified accounts and highly visible users usually play an important role in social media since their fact-checking tweets can reach many audiences [13, 27]. Since the verified accounts are more trustworthy, their fact-checking tweets are often shared by many other users. In our dataset, 2,401 guardians (2.2%) had verified accounts. Table 4 shows the top 15 verified accounts. Interestingly, some of these verified accounts behaved as D&S guardians, highlighted with the blue color in the table. Particularly, @PolitiFact, and @OpenSecretsDC, the official accounts of Politifact.com and OpenSecrets.org, frequently engaged in many online conversations. 8,221 guardians (7.5%) were highly visible users. Most top verified guardians, and many top S-guardians had a large number of followers. Altogether, S-tweets of the 45,406 S-guardians reached over 200 million followers.

Based on the analysis, we conclude that both D-guardians and S-guardians played important roles in terms of fact-checking claims and spreading the fact-checked news to the other users. Therefore, we need both types of guardians to spread credible information.

2.3 Temporal Behavior of Guardians in Fact-Checking Activities

To further understand activeness of guardians, we examined how quickly D-guardians posted their fact-checking URLs as responses to original posters' claims in online conversations. In particular, we measured response time of a D-tweet/D-guardian as a gap between an original poster's posting time and the fact-checking D-tweet's time. We collected all response time of D-tweets, grouped them and plotted a bar chart in Fig. 2a. The mean and median of response time were 2.26 days and 34 min, respectively. 90% of D-tweets were posted within one day, indicating D-guardians quickly responded to the claims and expressed their enthusiasm by posting fact-checking URLs/tweets.

Similarly, we also measured response time of an S-tweet/S-guardian (Fig. 2b) as a gap between D-tweet's posting time and the corresponding S-tweet's posting time. The mean and median of the response time were 3.1 days and 90 min, respectively. 88.5% of S-tweets were posted within 1 day, indicating S-guardians also quickly responded and spread fact-checked information.

Finally, we measured S-guardians' inter-posting time to understand how long it took between two consecutive S-tweets, given the corresponding D-tweet. First, we grouped S-tweets based on each corresponding D-tweet, and sorted them in the ascending order of S-tweet creation time. Next, within each group, we computed inter-posting time δ_i as a gap between two consecutive S-tweets i and $i + 1$ and created pairs of inter-posting time (δ_i, δ_{i+1}). These pairs were merged across all the groups and were plotted in log2 scale in Fig. 2c. Overall, the average inter-posting time was 5 min, which means an S-tweet was posted once per 5 min by S-guardians after the corresponding D-tweet was posted. To sum up, both D-guardians and S-guardians were active and quickly responded to claims and fact-checked content.

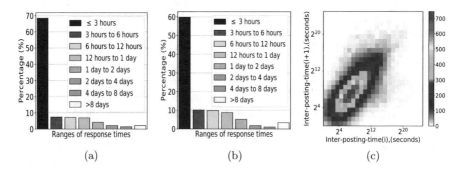

(a) (b) (c)

Fig. 2 Ranges of response time of D-guardians and S-guardians, and inter-posting time of S-tweets. The color in (**c**) indicates the number of pairs. (**a**) D-guardians' response time. (**b**) S-guardians' response time. (**c**) S-tweets' inter-posting time

2.4 Fact-Checking Article Recommendation Framework

In the previous section, we found that the guardians are highly active in fact-checking activities. To encourage them to further engage in disseminating fact-checked information, we propose a recommendation model to personalize fact-checking articles. The aim of the recommendation model is to help guardians quickly access new interesting fact-checking URLs/pages so that they could embed them in their messages, correct unverified claims or misinformation, and spread fact-checked information. We use terms "fact-checking URLs", "fact-checking articles" and "URL", interchangeably.

Problem Statement Let $\mathcal{N} = \{u_1, u_2, \ldots, u_N\}$ and $\mathcal{M} = \{\ell_1, \ell_2, \ldots, \ell_M\}$ be a set of N guardians and a set of M fact-checking URLs, respectively. We view the action of embedding a fact-checking URL ℓ_j into a fact-checking tweet of guardian u_i as an interaction pair (u_i, ℓ_j). We form a matrix $\mathbf{X} \in \mathbb{R}^{N \times M}$ where $\mathbf{X}_{ij} = 1$ if the guardian u_i posted a fact-checking URL ℓ_j. Otherwise, $\mathbf{X}_{ij} = 0$. Our main goal is to learn a model that recommends similar URLs to guardians whose interests are similar. In particular, we aim to learn matrix $\mathbf{U} \in \mathbb{R}^{N \times D}$, where each row vector $U_i^T \in \mathbb{R}^{D \times 1}$ is the latent representation of guardian u_i, and matrix $\mathbf{V} \in \mathbb{R}^{D \times M}$, where each column vector $V_j \in \mathbb{R}^{D \times 1}$ is the latent representation of URL ℓ_j. $D \ll min(M, N)$ is latent dimensions. Toward the goal, we propose our initial/basic matrix factorization model as follows:

$$\min_{\mathbf{U}, \mathbf{V}} \|\Omega \odot (\mathbf{X} - \mathbf{UV})\|_F^2 + \lambda(\|\mathbf{U}\|_F^2 + \|\mathbf{V}\|_F^2) \tag{1}$$

where $\Omega \in \mathbb{R}^{N \times M}$, and $\Omega_{ij} = 1$ if $\mathbf{X}_{ij} = 1$. Otherwise, $\Omega_{ij} = 0$. Operators $\odot$ and $\|.\|_F^2$ are Hadamard product and Frobenius norm, respectively. Finally, λ is regularization factor to avoid overfitting.

Co-ocurrence model Now, we turn to extend our basic model in Eq. 1 by further utilizing the interaction matrix $\mathbf{X}$. Inspired by [15, 20], we propose to regularize our basic model in Eq. 1 by generating two additional matrices – URL-URL co-occurrence matrix and guardian-guardian co-occurrence matrix. Our main intuition of the extension is that a pair of URLs, which were posted by the same guardian, may be similar to each other. Likewise, a pair of guardians who posted the same URLs may be alike. To better understand our proposed models, we present the word embedding model as background information.

Word embedding model Given a sequence of training words, word embedding models attempt to learn the distributed vector representation of each word. A typical example is *word2vec* proposed by Mikolov et al. [20]. Given a training word w, the main objective of the skip-gram model in *word2vec* is to predict the *context words* (i.e. the words that appear in a fixed-size context window) of w. Recently, it has been shown that training skip-gram model with negative sampling is similar to factorizing a word-context matrix named Shifted Positive Pointwise Mutual Information matrix

($SPPMI$) [14]. Given a word i and its context word j, the value $SPPMI(i, j)$ is computed as follows:

$$SPPMI(i, j) = max\{PMI(i, j) - log(s), 0\} \tag{2}$$

where $s \geq 1$ is the number of negative samples, and $PMI(i, j)$ is an element of Pointwise Mutual Information (PMI) matrix. $PMI(i, j)$ is estimated as $\log\left(\frac{\#(i,j) \times |D|}{\#(i) \times \#(j)}\right)$ where $\#(i, j)$ is the number of times that word j appears in the context window of word i. $\#(i) = \sum_j \#(i, j)$, and $\#(j) = \sum_i \#(i, j)$. $|D|$ is the total number of pairs of word and context word. Note that $PMI(i, i) = 0$ for every word i.

URL-URL co-occurrence We generate a matrix $\mathbf{R} \in \mathbb{R}^{M \times M}$ where $\mathbf{R}_{ij} = SPPMI(\ell_i, \ell_j)$ based on co-occurrence of URLs. In particular, for each URL ℓ_i posted by a specific guardian, we define its context as all other URLs ℓ_j posted by the same guardian. Based on this definition, $\#(i, j)$ means the number of guardians that posted both URL ℓ_i and ℓ_j. $\#(i, j)$ is also interpreted as the co-occurrence of URL ℓ_i and URL ℓ_j. After that, we compute $PMI(\ell_i, \ell_j)$ and $SPPMI(\ell_i, \ell_j)$ based on Eq. 2 for all pairs of ℓ_i and ℓ_j.

Guardian-Guardian co-occurrence Similarly, the context for each guardian u_i is defined as all other guardians u_j who posted the same URL with u_i. Then, $\#(i, j)$ is the number of URLs that both guardian u_i and guardian u_j commonly posted. Given this definition, we can generate a SPPMI matrix $\mathbf{G} \in \mathbb{R}^{N \times N}$ where $\mathbf{G}_{ij} = SPPMI(u_i, u_j)$. The same value of hyper-parameter s is used for generating matrices $\mathbf{R}$ and $\mathbf{G}$.

Regularizing matrix factorization with co-occurrence matrices Our intuition is that URLs which are commonly posted by similar set of guardians are similar, and guardians who commonly posted the same set of URLs are close to each other. With that intuition, we propose loss function $\mathcal{L}_{XRG}$ – a joint matrix factorization model of three matrices $\mathbf{X}$, $\mathbf{R}$ and $\mathbf{G}$ as follows:

$$\begin{aligned}
\mathcal{L}_{XRG} = \|\Omega \odot (\mathbf{X} - \mathbf{UV})\|_F^2 + \lambda(\|\mathbf{U}\|_F^2 + \|\mathbf{V}\|_F^2) \\
+ \|\mathbf{R}^{mask} \odot (\mathbf{R} - \mathbf{V}^T \mathbf{K})\|_F^2 + \|\mathbf{G}^{mask} \odot (\mathbf{G} - \mathbf{UL})\|_F^2
\end{aligned} \tag{3}$$

where $\mathbf{R}^{mask} \in \mathbb{R}^{M \times M}$, $\mathbf{R}_{ij}^{mask} = 1$ if $\mathbf{R}_{ij} > 0$. Otherwise, $\mathbf{R}_{ij}^{mask} = 0$. $\mathbf{G}^{mask} \in \mathbb{R}^{N \times N}$, $\mathbf{G}_{ij}^{mask} = 1$ if $\mathbf{G}_{ij} > 0$. Otherwise, $\mathbf{G}_{ij}^{mask} = 0$. Two matrices $\mathbf{K} \in \mathbb{R}^{D \times M}$ and $\mathbf{L} \in \mathbb{R}^{D \times N}$ act as additional parameters. Although our work shares similar ideas with [15], there are three key differences between our model and [15] as follows: (1) we omit bias matrices to reduce model complexity which is helpful in reducing overfitting, (2) additional matrix $\mathbf{G}$ is factorized and (3) we do not regularize parameters $\mathbf{K}$ and $\mathbf{L}$.

2.5 Integrating Auxiliary Information

In addition, we propose auxiliary information which will be integrated with Eq. 3 to improve URL recommendation performance.

Modeling social structure The social structure of guardians may reflect the homophily phenomenon indicating that guardians who follow each other may have similar interests in fact-checking URLs. To model this social structure of guardians, we first construct an unweighted undirected graph $G(V, E)$ where nodes are guardians, and an edge (u_i, u_j) between guardians u_i and u_j are formed if u_i follows u_j or u_j follows u_i. In our dataset, in total, there were 1,033,704 edges in $G(V, E)$ (density = 0.013898), which is 5.9 times higher than reported density in [31], indicating dense connections between guardians. We represent $G(V, E)$ by using an adjacency matrix $\mathbf{S} \in \mathbb{R}^{N \times N}$ where $\mathbf{S}_{ij} = 1$ if there is an edge (u_i, u_j). Otherwise, $\mathbf{S}_{ij} = 0$. Second, we use Eq. 4 as a regularization term to make latent representations of connected guardians similar to each other. Then, we formally minimize $\mathcal{L}_1$ as follows:

$$\mathcal{L}_1 = \|\mathbf{S} - \mathbf{U}\mathbf{U}^T\|_F^2 \qquad (4)$$

Modeling topical interests based on recent tweets In addition to social structure, the content of recent tweets may reflect guardians' interests [1, 2, 5]. For each guardian, we build a document by aggregating his/her 200 recent tweets and then employ the Doc2Vec model [12] to learn latent representations of the document. Doc2Vec is an unsupervised learning algorithm, which automatically learns high quality representation of documents. We use Gensim[5] as implementation of the Doc2Vec, set 300 as latent dimensions of documents, and train Doc2Vec model for 100 iterations. After training Doc2Vec model, we derive cosine similarity of every pair of learned vectors to create a symmetric matrix $\mathbf{X}_{uu} \in \mathbb{R}^{N \times N}$, where $\mathbf{X}_{uu}(i, j) \in [0; 1]$ represents the similarity of document vectors of guardians u_i and u_j. Intuitively, if two guardians have similar interests, their document vectors may be similar. Thus, we regularize guardians' latent representations to make them as close as possible by minimizing the following objective function:

$$
\begin{aligned}
\mathcal{L}_2 &= \frac{1}{2} \sum_{i=1, j=1}^{N} \mathbf{X}_{uu}(i, j) \|U_i^T - U_j^T\|^2 \\
&= \sum_{i=1}^{N} U_i^T \mathbf{D}_{uu}(i, i) U_i - \sum_{i=1, j=1}^{N} U_i^T \mathbf{X}_{uu}(i, j) U_j \\
&= Tr(\mathbf{U}^T \mathbf{D}_{uu} \mathbf{U}) - Tr(\mathbf{U}^T \mathbf{X}_{uu} \mathbf{U}) = Tr(\mathbf{U}^T \mathcal{L}_{uu} \mathbf{U})
\end{aligned}
\qquad (5)
$$

---

[5]https://radimrehurek.com/gensim/

where $\mathbf{D}_{uu} \in \mathbb{R}^{N \times N}$ is a diagonal matrix with elements on the diagonal $\mathbf{D}_{uu}(i,i) = \sum_{j=1}^{N} \mathbf{X}_{uu}(i,j)$. $Tr(.)$ is the trace of matrix, and $\mathcal{L}_{uu} = \mathbf{D}_{uu} - \mathbf{X}_{uu}$, which is a Laplacian matrix of the matrix $\mathbf{X}_{uu}$.

Modeling topical similarity of fact-checking pages We further exploit the content of fact-checking URLs (i.e., fact-checking pages) as an additional data source to improve recommendation quality. Intuitively, if the content of two URLs are similar, their latent representations should be close. Exploiting the content of a fact-checking URL has been employed in [2, 30]. In this paper,

---

Algorithm 1 GAU optimization algorithm

Input: Guardian-URL interaction matrix $\mathbf{X}$, URL-URL SPPMI matrix $\mathbf{R}$, Guardian-Guardian SPPMI matrix $\mathbf{G}$, social structure matrix $\mathbf{S}$, Laplacian matrix $\mathcal{L}_{uu}$ of guardians, Laplician matrix $\mathcal{L}_{\ell\ell}$ of URLs, binary matrices Ω, $\mathbf{R}^{mask}$ and $\mathbf{G}^{mask}$ as indication matrices.
Output: $\mathbf{U}$ and $\mathbf{V}$
1: Initialize $\mathbf{U}, \mathbf{V}, \mathbf{K}$ and $\mathbf{L}$ with Gaussian distribution $\mathcal{N}(0, 0.01^2)$, $t \leftarrow 0$
2: **while** Not Converged **do**
3:     Compute $\frac{\partial \mathcal{L}_{GAU}}{\partial \mathbf{U}}, \frac{\partial \mathcal{L}_{GAU}}{\partial \mathbf{V}}, \frac{\partial \mathcal{L}_{GAU}}{\partial \mathbf{L}}$ and $\frac{\partial \mathcal{L}_{GAU}}{\partial \mathbf{K}}$ in Eq. 9
4:     $\mathbf{U}_{t+1} \leftarrow \mathbf{U}_t - \eta \frac{\partial \mathcal{L}_{GAU}}{\partial \mathbf{U}}$
5:     $\mathbf{V}_{t+1} \leftarrow \mathbf{V}_t - \eta \frac{\partial \mathcal{L}_{GAU}}{\partial \mathbf{V}}$
6:     $\mathbf{L}_{t+1} \leftarrow \mathbf{L}_t - \eta \frac{\partial \mathcal{L}_{GAU}}{\partial \mathbf{L}}$
7:     $\mathbf{K}_{t+1} \leftarrow \mathbf{K}_t - \eta \frac{\partial \mathcal{L}_{GAU}}{\partial \mathbf{K}}$
8:     $t \leftarrow t+1$
    return $\mathbf{U}$ and $\mathbf{V}$

---

we apply a different approach, in which the Doc2Vec model is utilized to learn latent representation of URLs. Hyperparameters of the Doc2Vec model are the same as what we used for content of tweets. After training the Doc2Vec model, we derive the symmetric similarity matrix $\mathbf{X}_{\ell\ell} \in \mathbb{R}^{M \times M}$ and minimize the loss function $\mathcal{L}_3$ in Eq. 6 as a way to regulate latent representation of URLs.

$$\mathcal{L}_3 = \frac{1}{2} \sum_{i=1, j=1}^{M} \mathbf{X}_{\ell\ell}(i,j) \| V_i - V_j \|^2$$

$$= \sum_{i=1}^{M} V_i \mathbf{D}_{\ell\ell}(i,i) V_i^T - \sum_{i=1, j=1}^{M} V_i \mathbf{X}_{\ell\ell}(i,j) V_j^T \qquad (6)$$

$$= Tr(\mathbf{V}(\mathbf{D}_{\ell\ell} - \mathbf{X}_{\ell\ell})\mathbf{V}^T)$$

$$= Tr(\mathbf{V}\mathcal{L}_{\ell\ell}\mathbf{V}^T)$$

where $\mathbf{D}_{\ell\ell} \in \mathbb{R}^{M \times M}$ is a diagonal matrix with $\mathbf{D}_{\ell\ell}(i,i) = \sum_{j=1}^{M} \mathbf{X}_{\ell\ell}(i,j)$ and $\mathcal{L}_{\ell\ell} = \mathbf{D}_{\ell\ell} - \mathbf{X}_{\ell\ell}$, which is the graph Laplacian of the matrix $\mathbf{X}_{\ell\ell}$.

2.6 Joint-Learning Fact-Checking URL Recommendation Model

Finally, we propose GAU – a joint model of **Guardian**-Guardian SPPMI matrix, **A**uxiliary information and **U**RL-URL SPPMI matrix. The objective function of our model, $\mathcal{L}_{GAU}$, is presented in Eq. 7:

$$\min_{\mathbf{U},\mathbf{V},\mathbf{L},\mathbf{K}} \mathcal{L}_{GAU} = \|\Omega \odot (\mathbf{X} - \mathbf{UV})\|_F^2 + \lambda(\|\mathbf{U}\|_F^2 + \|\mathbf{V}\|_F^2)$$

$$+ \|\mathbf{R}^{mask} \odot (\mathbf{R} - \mathbf{V}^T\mathbf{K})\|_F^2$$

$$+ \|\mathbf{G}^{mask} \odot (\mathbf{G} - \mathbf{UL})\|_F^2 \qquad (7)$$

$$+ \alpha \times \|\mathbf{S} - \mathbf{UU}^T\|_F^2$$

$$+ \gamma \times Tr(\mathbf{U}^T \mathcal{L}_{uu}\mathbf{U})$$

$$+ \beta \times Tr(\mathbf{V}\mathcal{L}_{\ell\ell}\mathbf{V}^T)$$

where $\alpha, \gamma, \beta, \lambda$ and shifted negative sampling value s are hyper parameters, tuned based on a validation set. We optimize $\mathcal{L}_{GAU}$ by using gradient descent to iteratively update parameters with fixed learning rate $\eta = 0.001$. The details of the optimization algorithm are presented in Algorithm 1. After learning $\mathbf{U}$ and $\mathbf{V}$, we estimate the guardian u_i's preference for URL ℓ_j as: $\hat{r}_{i,j} \approx U_i V_j$. The final URLs recommended for a guardian u_i is formed based on ranking:

$$u_i : \ell_{j_1} > \ell_{j_2} > \ldots > \ell_{j_M} \rightarrow \hat{r}_{i,j_1} > \hat{r}_{i,j_2} > \ldots > \hat{r}_{i,j_M} \qquad (8)$$

The derivatives of loss $\mathcal{L}_{GAU}$ with respect to parameters $\mathbf{U}$, $\mathbf{V}$, $\mathbf{K}$ and $\mathbf{L}$ are:

$$\frac{\partial \mathcal{L}_{GAU}}{\partial \mathbf{U}} = -2(\Omega \odot \Omega \odot (\mathbf{X} - \mathbf{UV}))\mathbf{V}^T + 2\lambda \times (\mathbf{U})$$

$$-2(\mathbf{G}^{mask} \odot \mathbf{G}^{mask} \odot (\mathbf{G} - \mathbf{UL}))\mathbf{L}^T$$

$$-2\alpha((\mathbf{S} - \mathbf{UU}^T + (\mathbf{S} - \mathbf{UU}^T)^T)\mathbf{U})$$

$$+\gamma \times (\mathcal{L}_{uu} + \mathcal{L}_{uu}^T)\mathbf{U}$$

$$\frac{\partial \mathcal{L}_{GAU}}{\partial \mathbf{V}} = -2\mathbf{U}^T(\Omega \odot \Omega \odot (\mathbf{X} - \mathbf{UV})) + 2\lambda \times (\mathbf{V}) \qquad (9)$$

$$-2\mathbf{K}(\mathbf{R}^{mask} \odot \mathbf{R}^{mask} \odot (\mathbf{R} - \mathbf{V}^T\mathbf{K}))^T$$

$$+\beta \times \mathbf{V}(\mathcal{L}_{\ell\ell} + \mathcal{L}_{\ell\ell}^T)$$

$$\frac{\partial \mathcal{L}_{GAU}}{\partial \mathbf{L}} = -2\mathbf{U}^T(\mathbf{G}^{mask} \odot \mathbf{G}^{mask} \odot (\mathbf{G} - \mathbf{UL}))$$

$$\frac{\partial \mathcal{L}_{GAU}}{\partial \mathbf{K}} = -2\mathbf{V}(\mathbf{R}^{mask} \odot \mathbf{R}^{mask} \odot (\mathbf{R} - \mathbf{V}^T\mathbf{K}))$$

2.7   Experimental Design and Evaluation Metrics

We were interested in selecting active and professional guardians who frequently posted fact-checking URLs since they would be more likely to spread recommended fact-checking URLs than casual guardians. We only selected guardians who used at least three distinct fact-checking URLs in their D-tweets and/or S-tweets. Altogether, 12,197 guardians were selected for training and evaluating recommendation models. They posted 4,834 distinct fact-checking URLs in total. The number of interactions was 68,684 (Sparsity:99.9%). There were 9,710 D-guardians, 6,674 S-guardians and 4,187 users who played both roles. The total number of followers of the 12,197 guardians was 55,325,364, indicating their high impact on fact-checked information propagation.

To validate our model, we randomly selected 70%, 10% and 20% URLs of each guardian for training, validation and testing. The validation data was used to tune hyper-parameters and to avoid overfitting. We repeated this evaluation scheme for five times, getting five different sets of training, validation and test data. The average results were reported. We used three standard ranking metrics such as Recall@k, MAP@k (Mean Average Precision) and NDCG@k (Normalized Discounted Cumulative Gain). We tested our model with $k \in \{5, 10, 15\}$.

2.8   Effectiveness of Auxiliary Information and SPPMI Matrices

Before comparing our GAU model with four baselines, which will be described in the following section, we first examined the effectiveness of exploiting auxiliary information and the utility of jointly factorizing SPPMI matrices. Starting from our basic model in Eq. 1, we created variants of the GAU model. Since there are many variants of GAU, we selectively report performance of the following GAU's variants:

- Our basic model (Eq. 1) (BASIC)
- BASIC + Network + URL's content (BASIC + NW + UC)
- BASIC + Network + URL's content + URL's SPPMI matrix (BASIC + NW + UC + SU)
- BASIC + URL's SPPMI matrix + Guardians' SPPMI matrix (BASIC + SU + SG)
- BASIC + Network + URL's content + SPPMI matrix of URLs + SPPMI matrix of Guardians (BASIC + NW + UC + SU + SG)
- Our GAU model

Table 5 shows performance of the variants and the GAU model. It shows the rank of each method based on the reported metrics. By adding social network information and fact-checking URL's content to Eq. 1, there was a huge climb in

Table 5 Effectiveness of using auxiliary information and co-occurrence matrices. The GAU model outperforms the other variants significantly with p-value <0.001

Methods	Recall@5	NDCG@5	MAP@5	Recall@10	NDCG@10	MAP@10	Recall@15	NDCG@15	MAP@15	Avg. Rank
BASIC	0.089 (6)	0.060 (6)	0.048 (6)	0.132 (6)	0.074 (6)	0.054 (6)	0.162 (6)	0.082 (6)	0.056 (6)	6.0
BASIC+NW+UC	0.099 (4)	0.068 (5)	0.055 (5)	0.148 (4)	0.084 (4)	0.062 (5)	0.183 (3)	0.093 (5)	0.064 (5)	4.4
BASIC+NW+UC+SU	0.099 (5)	0.068 (4)	0.056 (4)	0.147 (5)	0.084 (5)	0.062 (4)	0.183 (4)	0.094 (4)	0.065 (4)	4.3
BASIC+SU+SG	0.102 (3)	0.069 (3)	0.057 (3)	0.149 (3)	0.085 (3)	0.063 (3)	0.182 (5)	0.094 (3)	0.066 (3)	3.2
BASIC+NW+UC+SU+SG	0.111 (2)	0.074 (2)	0.060 (2)	0.161 (2)	0.091 (2)	0.066 (2)	0.195 (2)	0.099 (2)	0.069 (2)	2.0
Our GAU model	0.116 (1)	0.079 (1)	0.065 (1)	0.164 (1)	0.095 (1)	0.071 (1)	0.197 (1)	0.104 (1)	0.074 (1)	1.0

performance of BASIC+NW+UC over BASIC across all metrics. In particular, Recall, NDCG and MAP of BASIC+NW+UC were better than BASIC about $12.20\% \pm 1.31\%$, $13.39\% \pm 0.34\%$ and $14.04\% \pm 0.76\%$, respectively (confidence interval 95%). These results confirm the effectiveness of exploiting the auxiliary information.

How about using co-occurrence SPPMI matrices of fact-checking URLs and guardians? First, when adding co-occurrence SPPMI matrix of fact-checking URL (SU) to the variant BASIC+NW+UC, we did not see much improvement across all settings. Second, when jointly factorizing two SPPMI matrices (BASIC+SU+SG) and comparing it with the variant BASIC+NW+UC, we can see that BASIC+SU+SG and BASIC+NW+UC performed equally well. Again, BASIC+SU+SG did not use any additional data sources except the interaction matrix **X**. It is an attractive benefit since it did not depend on other data sources. In other words, it reflects that regularizing the BASIC model with SPPMI matrices is comparable to adding network data and URLs' contents to the BASIC model.

So far, both auxiliary information and SPPMI matrices are beneficial to improving recommendation quality. How about combining all of them into a single model? Will performance be further improved? We turned to the variant BASIC+NW+UC+SU+SG. As expected, BASIC+NW+UC+SU+SG enhanced SU+SG by $7.90\% \pm 1.79\%$ Recall, $6.58\% \pm 0.40\%$ NDCG, and $5.53\% \pm 0.22\%$ MAP. Its results were also higher than BASIC+NW+UC about $9.10\% \pm 6.15\%$ Recall, $7.92\% \pm 2.50\%$ NDCG and $7.75\% \pm 0.58\%$ MAP.

Since adding auxiliary data was valuable, we now exploit another data source – 200 recent tweets' content. Consistently, adding the tweets' content indeed improved performance. The improvement of GAU over BASIC+NW+UC+SU+SG model was 4.0% Recall, 6.6% NDCG and 8.4% MAP. This improvement is statistically significant with p-value<0.001 using Wilcoxon one-sided test. Comparing the GAU with the BASIC model, we observed a dramatic increase in performance across all metrics. Specifically, Recall, NDCG and MAP were improved by $25.13\% \pm 10.64\%$, $28.64\% \pm 7.13\%$ and $32\% \pm 4.29\%$ respectively.

Based on the experiments, we conclude that the auxiliary data as well as co-occurrence matrices are helpful to improve recommendation quality. Adding SU+SG or NW+UC enhanced the BASIC model by 12–14%. Our GAU model performed best, improving 25~32% compared with the BASIC model.

2.9   Performance of Our Model and Baselines

We compared our proposed model with the following four state-of-the-art collaborative filtering algorithms:

– **BPRMF** Bayesian Personalized Ranking Matrix Factorization [23] optimizes the matrix factorization model with pairwise ranking loss. It is a common baseline for item recommendation.

- **MF** Matrix Factorization (MF) [10] is a standard technique in collaborative filtering. Given an interaction matrix $\mathbf{X} \in \mathbb{R}^{N \times M}$, it factorizes $\mathbf{X}$ into two matrices $\mathbf{U} \in \mathbb{R}^{N \times D}$ and $\mathbf{V} \in \mathbb{R}^{D \times M}$, which are latent representations of users and items, respectively.
- **CoFactor** CoFactor [15] extended Weighted Matrix Factorization (WMF) by jointly decomposing interaction matrix $\mathbf{X}$ and co-occurrence SPPMI matrix for items (i.e., fact-checking URLs in this context). We set a confidence value $c_{X_{ij}=1} = 1.0$ for $X_{ij} = 1$, and we set $c_{X_{ij}=0} = 0.01$ for non-observed interaction. The number of negative samples s was grid-searched in a set $s \in \{1, 2, 5, 10, 50\}$, following the same settings as in [15].
- **CTR** Collaborative Filtering Regression [30] employed content of URLs (i.e., fact-checking pages in this context) to recommend scientific papers to users. Following exactly the best setting reported in the paper, we selected the top 8,000 words from fact-checking URLs' contents based on the mean of tf-idf values and set $\lambda_u = 0.01$, $\lambda_v = 100$, $D = 200$, $a = 1$ and $b = 0.01$.

To build our GAU model, we conducted the grid-search to select the best value of α, β and γ in $\{0.02, 0.04, 0.06, 0.08\}$. The number of negative samples s for constructing SPPMI matrices was in $\{1, 2, 5, 10, 50\}$. For all of the baselines and the GAU model, we set latent dimensions to $D = 100$ unless explicitly stated, and regularization value λ was grid-searched in $\{10^{-5}, 3 \times 10^{-5}, 5 \times 10^{-5}, 7 \times 10^{-5}\}$ by default. We only report the best result of each baseline.

Figure 3 shows the performance of the four baselines and GAU. MF was better than BPRMF which was designed to optimize Area Under Curve (AUC). CTR was a very competitive baseline. This reflects the importance of fact-checking URL's content (i.e., fact-checking page) in recommending right fact-checking URLs to guardians. GAU performed better than CTR by 12.75% $\pm$ 0.95% Recall, 11.2% $\pm$ 4.6% NDCG, and 12.5% $\pm$ 2.5% MAP. GAU also outperformed CoFactor with a large margin by 25.8% $\pm$ 8.4% Recall, 29.2% $\pm$ 5.8% NDCG, and 32.6% $\pm$ 3.4% MAP (confidence interval 95%). Overall, our GAU model significantly outperformed all the baselines (p-value<0.001). The improvement over the baselines was 11~33%.

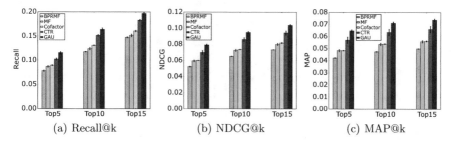

(a) Recall@k   (b) NDCG@k   (c) MAP@k

Fig. 3 Performance of our GAU model and 4 baselines. The GAU model outperforms the baselines (p-value < 0.001). (**a**) Recall@k. (**b**) NDCG@k. (**c**) MAP@k

2.10 Exploiting Hyper-Parameters

We investigated the impact of hyper-parameters α, β and γ on the GAU model. These hyper-parameters control the contribution of social network, fact-checking URL's content and 200 recent tweets' content to the GAU. We tested α, β and γ from 0.01 to 0.09, increasing 0.01 in each step, and then report the average recall@15, while we fixed $\lambda = 3 \times 10^{-5}$ and the number of negative samples $s = 10$. In Fig. 4a, we fixed $\beta = 0.08$ and varied α and γ. The general trend was that recall@15 gradually went up, when α and γ increased. It reached the peak, when $\alpha = 0.06$ and $\gamma = 0.06$. Next, we fixed $\alpha = 0.08$. It seems recall@15 fluctuated when varying β and γ, but the amplitude was small. The max Recall@15 was only 2.2% larger than the smallest Recall@15. Finally, γ was fixed to 0.08. The trend was similar to Fig. 4a. In general, when α, β and γ are large, the performance tends to improve, which suggests the importance of regularizing our model using the auxiliary information.

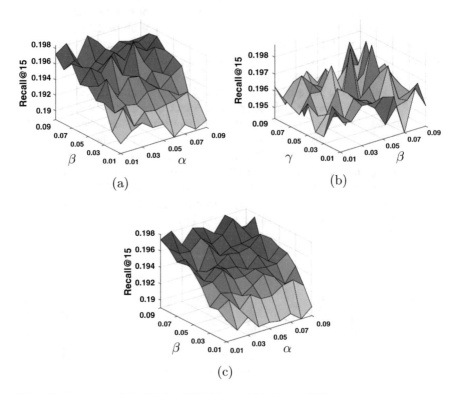

Fig. 4 Parameter sensitivity. (**a**) $\beta = 0.08$. (**b**) $\alpha = 0.08$. (**c**) $\gamma = 0.08$

2.11 Discussion

So far, we identified who guardians are and their temporal behavior. Although there are highly active guardians in fact-checking guardians, most guardians only posted 1~2 fact-checking tweets. Therefore, we only target active guardians, who posted at least 3 fact-checking URLs since guardians may continue to be active in spreading fact-checked information in the future. Another observation is that the top verified guardians seem not to be active in the covered time period. This phenomenon may be explained by the fact that these verified guardians may be cautious about what they should post to their followers. From our experiments, we showed that integrating auxiliary information is useful in improving recommendation quality. Although our model outperforms baselines, there are considerable space to improve our model. For example, we may utilize contents of original tweets, temporal factors and activeness of guardians. Deep Learning architectures may help us improve our model. We leave these directions for future exploration.

3 Fact-Checking Responses Generation Framework

In this section, we turn our attention to generating responses with fact-checking intention to help guardians fact-check information faster and as a result increase their engagement in fact-checking activities. Since S-tweets are mostly copies of Direct Fact-checking tweets (D-tweets), we focus on generating D-tweets when an original tweet is given.

3.1 Datasets of Original Tweets and Direct Fact-Checking Tweets

Since training an effective text generation framework requires large number of pairs of original tweets and D-tweets, we extend our dataset in Sect. 2.1 with additional D-tweets collected from Hoaxy system. Totally, we collected 247,436 distinct D-tweets posted between May 16, 2016 and May 26, 2018. We removed non-English D-tweets, and D-tweets containing fact-checking URLs linked to non-article pages such as the main page and about page of a fact-checking site. Then, among the remaining D-tweets, if its corresponding original tweet was deleted or was not accessible via Twitter APIs because of suspension of an original poster, we further filtered out the D-tweets. As a result, 190,158 D-tweets and 164,477 distinct original tweets were remained.

To further ensure that each of the remaining D-tweets reflected fact-checking intention and make a high quality dataset, we only kept a D-tweet whose fact-checking article was rated as true or false. Our manual verification of 100 random samples confirmed that D-tweets citing fact-checking articles with true or false label contained clearer fact-checking intention than D-tweets with other labels such as half true or mixture. In other words, D-tweets associated with mixed labels were discarded. After the pre-processing steps, our final dataset consisted of 73,203 D-tweets and 64,110 original tweets posted by 41,732 distinct D-guardians, and 44,411 distinct original posters, respectively. We use this dataset in the following sections.

3.2 Response Generation Framework

Formally, given a pair of an original tweet and a D-tweet, the original tweet x is a sequence of words $x = \{x_i | i \in [1; N]\}$ and the D-tweet is another sequence of words $y = \{y_j | j \in [1; M]\}$, where N and M are the length of the original tweet and the length of D-tweet, respectively. We inserted a special token <s> as a starting token into every D-tweet. Drawing inspiration from [18], we propose and build a framework as shown in Fig. 5 that consists of three main components: (i) the shared word embedding layer, (ii) the encoder to capture representation of the original tweet and (iii) the decoder to generate a D-tweet. Their details are as follows:

Shared Word Embedding Layer For every word x_i in the original tweet x, we represent it as a one-hot encoding vector $x_i \in \mathbb{R}^V$ and embed it into a D-dimensional vector $\mathbf{x}_i \in \mathbb{R}^D$ as follows: $\mathbf{x}_i = \mathbf{W}_e x_i$, where $\mathbf{W}_e \in \mathbb{R}^{D \times V}$ is an embedding matrix and V is the vocabulary size. We use the same word embedding matrix $\mathbf{W}_e$ for the D-tweet. In particular, for every word y_i (represented as one-hot vector $y_i \in \mathbb{R}^V$) in the D-tweet y, we embed it into a vector $\mathbf{y}_i = \mathbf{W}_e y_i$. The embedding matrix $\mathbf{W}_e$ is a learned parameter and could be initialized by either pre-trained word vectors (e.g.

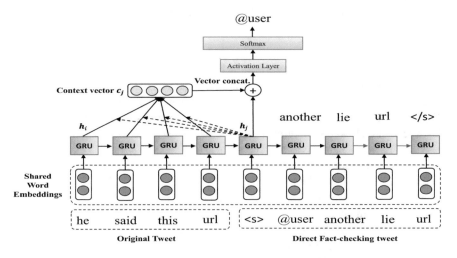

Fig. 5 Our proposed framework to generate responses with fact-checking intention

Glove vectors) or random initialization. Since our model is designed specifically for fact-checking domain, we initialized $\mathbf{W}_e$ with Normal Distribution $\mathcal{N}(0, 1)$ and trained it from scratch. By using a shared $\mathbf{W}_e$, we could reduce the number of learned parameters significantly compared with [18].

Encoder The encoder is used to learn latent representation of the original tweet x. We adopt a Recurrent Neural Network (RNN) to represent the encoder due to its large capacity to condition each word x_i on all previous words $x_{<i}$ in the original tweet x. To overcome the vanishing or exploding gradient problem of RNN, we choose Gated Recurrent Unit (GRU) [6]. Formally, we compute hidden state $\mathbf{h}_i \in \mathbb{R}^H$ at time-step ith in the encoder as follows:

$$\mathbf{h}_i = GRU(\mathbf{x}_i, \mathbf{h}_{i-1}) \tag{10}$$

where the GRU is defined by the following equations:

$$
\begin{aligned}
\mathbf{z}_i &= \sigma(\mathbf{x}_i \mathbf{W}_z + \mathbf{h}_{i-1} \mathbf{U}_z) \\
\mathbf{r}_i &= \sigma(\mathbf{x}_i \mathbf{W}_r + \mathbf{h}_{i-1} \mathbf{U}_r) \\
\tilde{\mathbf{h}}_i &= tanh(\mathbf{x}_i \mathbf{W}_o + (\mathbf{r}_i \odot \mathbf{h}_{i-1}) \mathbf{U}_o) \\
\mathbf{h}_i &= (1 - \mathbf{z}_i) \odot \tilde{\mathbf{h}}_i + \mathbf{z}_i \odot \mathbf{h}_{i-1}
\end{aligned}
\tag{11}
$$

where $\mathbf{W}_{[z,r,o]}, \mathbf{U}_{[z,r,o]}$ are learned parameters. $\tilde{\mathbf{h}}_i$ is the new updated hidden state, $\mathbf{z}_i$ is the update gate, $\mathbf{r}_i$ is the reset gate, $\sigma(.)$ is the sigmoid function, $\odot$ is element wise product, and $\mathbf{h}_0 = \mathbf{0}$. After going through every word of the original tweet x, we have hidden states for every time-step $\mathbf{X} = [\mathbf{h}_1 \oplus \mathbf{h}_2 \oplus \ldots \oplus \mathbf{h}_N] \in \mathbb{R}^{H \times N}$, where $\oplus$ denotes concatenation of hidden states. We use the last hidden state $\mathbf{h}_N$ as features of the original tweet $\mathbf{x} = \mathbf{h}_N$.

Decoder The decoder takes $\mathbf{x}$ as the input to start the generation of a D-tweet. We use another GRU to represent the decoder to generate a sequence of tokens $y = \{y_1, y_2, \ldots, y_M\}$. At each time-step jth, the hidden state h_j is computed by another GRU: $\mathbf{h}_j = GRU(\mathbf{y}_j, \mathbf{h}_{j-1})$ where initial hidden states are $\mathbf{h}_0 = \mathbf{x}$. To provide additional context information when generating word y_j, we apply an attention mechanism to learn a weighted interpolation context vector $\mathbf{c}_j$ dependent on all of the hidden states output from all time-steps of the encoder. We compute $\mathbf{c}_j = \mathbf{X} \mathbf{a}_j$ where each component a_{ji} of $\mathbf{a}_j \in \mathbb{R}^N$ is the alignment score between the jth word in the D-tweet and the ith output from the encoder. In this study, $\mathbf{a}_j$ is computed by one of the following ways:

$$
\mathbf{a}_j =
\begin{cases}
softmax(\mathbf{X}^T \mathbf{h}_j) & \text{Dot Attention} \\
softmax(\mathbf{X}^T \mathbf{W}_a \mathbf{h}_j) & \text{Bilinear Attention}
\end{cases}
\tag{12}
$$

where softmax(.) is a softmax activation function and $\mathbf{W}_a \in \mathbb{R}^{H \times H}$ is a learned weight matrix. Note that we tried to employ other attention mechanisms including additive attention [3] and concat attention [18] but the above attention mechanisms in Eq. 12 produced better results. After computing the context vector $\mathbf{c}_j$, we concatenate $\mathbf{h}_j^T$ with $\mathbf{c}_j^T$ to obtain a richer representation. The word at jth time-step is predicted by a softmax classifier:

$$\hat{\mathbf{y}}_j = softmax\left(\mathbf{W}_s \tanh\left(\mathbf{W}_c[\mathbf{c}_j^T \oplus \mathbf{h}_j^T]^T\right)\right) \tag{13}$$

where $\mathbf{W}_c \in \mathbb{R}^{O \times 2H}$, and $\mathbf{W}_s \in \mathbb{R}^{V \times O}$ are weight matrices of a two-layer feedforward neural network and O is the output size. $\hat{\mathbf{y}}_j \in \mathbb{R}^V$ is a probability distribution over the vocabulary. The probability of choosing word v_k in the vocabulary as output is:

$$p(y_j = v_k|y_{j-1}, y_{j-2}, \ldots, y_1, \mathbf{x}) = \hat{\mathbf{y}}_{jk} \tag{14}$$

Therefore, the overall probability of generating the D-tweet y given the original tweet x is computed as follows:

$$p(y|x) = \prod_{j=1}^{M} p(y_j|y_{j-1}, y_{j-2}, \ldots, y_1, \mathbf{x}) \tag{15}$$

Since the entire architecture is differentiable, we jointly train the whole network with Teacher Forcing via Adam optimizer by minimizing the negative conditional log-likelihood for m pairs of the original tweet $x^{(i)}$ and the D-tweet $y^{(i)}$ as follows:

$$\min_{\theta_e, \theta_d} \mathcal{L} = -\sum_{i=1}^{m} \log p(y^{(i)}|x^{(i)}; \theta_e, \theta_d) \tag{16}$$

where θ_e and θ_d are the parameters of the encoder and the decoder, respectively. At test time, we used beam search to select top K generated responses. The generation process of a D-tweet is ended when an end-of-sentence token (e.g. </s>) is emitted.

3.3   Evaluation

In this section, we thoroughly evaluate our models namely **FCRG-DT** (based on dot attention in Eq. 12) and **FCRG-BL** (based on bilinear attention in Eq. 12) quantitatively and qualitatively. Since our methods are deterministic models, we compare them with state-of-the-art baselines in this direction.

- **SeqAttB:** Shang et al. [25] proposed a hybrid model that combines global scheme and local scheme [3] to generate responses for original tweets on Sina

Weibo. This model is one of the first work that generate responses for short text conversations.

- **HRED:** It [24] employs hierarchical RNNs for capturing information in a long context. HRED is a competitive method and a commonly used baseline for dialog generation systems.
- **our FCRG-BL:** This model uses the bilinear attention.
- **our FCRG-DT:** This model uses the dot attention.

Data Processing Similar to [24] in terms of text generation, we replaced numbers with `<number>` and personal names with `<person>`. Words that appeared less than three times were replaced by `<unk>` token to further mitigate the sparsity issue. Our vocabulary size was 15,321. The min, max and mean |tokens| of the original tweets were 1, 89 and 19.1, respectively. The min, max and mean |tokens| of D-tweets were 3, 64 and 12.3, respectively. Only 791 (1.2%) original tweets contained 1 token which is mostly a URL.

Experimental Design We randomly divided 73,203 pairs of the original tweets and D-tweets into training/validation/test sets with a ratio of 80%/10%/10%, respectively. The validation set was used to tune hyperparameters and for early stopping. At test time, we used the beam search to generate 15 responses per original tweet (beam size=15), and report the average results. To select the best hyperparameters, we conducted the standard grid search to choose the best value of a hidden size $H \in \{200, 300, 400\}$, and an output size $O \in \{256, 512\}$. We set word embedding size D to 300 by default unless explicitly stated. The length of the original tweets and D-tweets were set to the maximum value $N = 89$ and $M = 64$, respectively. The dropout rate was 0.2. We used Adam optimizer with fixed learning rate $\lambda = 0.001$, batch size $b = 32$, and gradient clipping was 0.25 to avoid exploded gradient. The same settings are applied to all models for the fair comparison.

A well known problem of the RNN-based decoder is that it tends to generate short responses. In our domain, examples of commonly generated responses were *fake news url.*, *you lie url.*, and *wrong url.* Because a very short response may be less interesting and has less power to be shared, we forced the beam search to generate responses with at least τ tokens. Since 92.4% of D-tweets had |tokens| $\geq$ 5, and 60% D-tweets had |tokens| $\geq$ 10, we chose $\tau \in \{0, 5, 10\}$. In practice, fact-checkers can choose their preferred |tokens| of generated responses by varying τ.

Evaluation Metrics To measure performance of our models and baselines, we adopted several syntactic and semantic evaluation metrics used in the prior works. In particular, we used word overlap-based metrics such as BLEU scores [22], ROUGE-L [16], and METEOR [4]. These metrics evaluate the amount of overlapping words between a generated response and a ground-truth D-tweet. The higher score indicates that the generated response are close/similar to the ground-truth D-tweet syntactically. In other words, the generated response and the D-tweet have a large number of overlapping words. Additionally, we also used embedding metrics (i.e. Greedy Matching and Vector Extrema) [17]. These metrics usually estimate sentence-level vectors by using some heuristic to combine the individual word vectors in the sentence. The sentence-level vectors between a generated response and the ground-truth D-tweet are compared by a measure such as cosine similarity. The higher value means the response and the D-tweet are semantically similar.

Quantitative Results Based on Word Overlap-Based Metrics In this experiment, we quantitatively measure performances of all models by using BLEU, ROUGE-L, and METEOR. Table 6 shows results in the test set. Firstly, our FCRG-DT and FCRG-BL performed equally well, and outperformed the baselines – SeqAttB and HRED. In practice, FCRG-DT model is more preferable due to fewer parameters compared with FCRG-BL. Overall, our models outperformed SeqAttB perhaps because fusing global scheme (i.e. the last hidden state of the encoder) and output hidden state of every time-step ith in the encoder may be less effective than using only the latter one to compute context vector c_j. HRED model utilized only global context without using context vector c_j in generating responses, leading to suboptimal results compared with our models.

Under no constraints on |tokens| of generated responses, our FCRG-DT achieved 6.24% ($p < 0.001$) improvement against SeqAttB on BLEU-3 according to Wilcoxon one-sided test. In BLEU-4, FCRG-DT improved SeqAttB by 7.32% and HRED by 7.76% ($p < 0.001$). In ROUGE-L, FCRG-DT improved SeqAttB and HRED by 3.32% and 4.31% with $p < 0.001$, respectively. In METEOR, our FCRG-DT and FCRG-BL achieved comparable performance with the baselines.

When |tokens| ≥ 5, we even achieve better results. The improvements of FCRG-DT over SeqAttB were 7.05% BLEU-3, 7.37% BLEU-4 and 3.25% ROUGE-L ($p < 0.001$). In comparison with HRED, the improvements of FCRG-DT were 5.25% BLEU-3, 5.64% BLEU-4, and 2.97% ROUGE-L ($p < 0.001$). Again, FCRG-DT are comparable with SeqAttB and HRED in METEOR measurement.

When |tokens| ≥ 10, there was a decreasing trend across metrics as shown in Table 6. It makes sense because generating longer response similar with a ground-truth D-tweet is much harder problem. Therefore, in reality, the Android messaging service recommends a very short reply (e.g., okay, yes, I am indeed) to reduce inaccurate risk. Despite the decreasing trend, our FCRG-DT and FCRG-BL improved the baselines by a larger margin. In particular, in BLEU-3, FCRG-DT outperformed SeqAttB and HRED by 17.9% and 16.0% ($p < 0.001$), respectively. For BLEU-4, the improvements of FCRG-DT over SeqAttB and HRED were 13.02% and 11.74% ($p < 0.001$), respectively. We observed consistent improvements over the baselines in ROUGE-L and METEOR. Overall, our models outperformed the baselines in terms of all of the word overlap-based metrics.

Quantitative Results Based on Embedding Metrics We adopted two embedding metrics to measure semantic similarity between generated responses and ground-truth D-tweets [17]. Again, we tested all the models under three settings as shown in Table 6. Our FCRG-DT performed best in all embedding metrics. Specifically, FCRG-DT outperformed SeqAttB by 3.98% and HRED by 6.00% improvements with $p < 0.001$ in Greedy Matching. FCRG-DT's improvements over SeqAttB and HRED were 26.24% and 5.62% ($p < 0.001$), respectively in Vector Extrema. When |tokens| ≥ 5, our FCRG-DT also outperformed the baselines in both Greedy

Table 6 Performance of our models and baselines. Our models outperformed baselines with p-value < 0.001

τ	Model	BLEU-2	BLEU-3	BLEU-4	ROUGE-L	METEOR	Greedy Mat.	Vector Ext.	Avg. Rank
τ = 0	SeqAttB	7.15 (4)	4.05 (4)	3.26 (3)	26.47 (3)	17.66 (3)	43.57 (3)	15.84 (4)	3.43
	HRED	7.30 (3)	4.07 (3)	3.25 (4)	26.22 (4)	17.55 (4)	42.73 (4)	18.93 (3)	3.57
	FCRG-BL	**7.68** (1)	4.27 (2)	3.41 (2)	27.14 (2)	**17.87** (1)	43.71 (2)	**20.24** (1)	1.57
	FCRG-DT	7.64 (2)	**4.30** (1)	**3.50** (1)	**27.35** (1)	17.75 (2)	**45.30** (1)	19.99 (2)	1.43
τ = 5	SeqAttB	7.47 (4)	4.09 (4)	3.18 (4)	26.17 (4)	17.72 (3)	41.04 (4)	14.69 (4)	3.86
	HRED	7.63 (3)	4.16 (3)	3.23 (3)	26.24 (3)	17.62 (4)	41.93 (3)	18.85 (3)	3.14
	FCRG-BL	7.93 (2)	4.29 (2)	3.30 (2)	26.95 (2)	**17.89** (1)	42.90 (2)	**20.05** (1)	1.71
	FCRG-DT	**8.04** (1)	**4.37** (1)	**3.41** (1)	**27.02** (1)	17.77 (2)	**44.38** (1)	19.44 (2)	1.29
τ = 10	SeqAttB	6.40 (4)	3.32 (4)	2.43 (4)	22.25 (4)	16.57 (4)	36.29 (4)	10.20 (4)	4.00
	HRED	6.54 (3)	3.37 (3)	2.46 (3)	22.98 (3)	17.11 (3)	37.51 (3)	15.54 (3)	3.00
	FCRG-BL	7.58 (2)	3.78 (2)	2.66 (2)	**25.09** (1)	**17.83** (1)	**39.81** (1)	**17.61** (1)	1.43
	FCRG-DT	**7.96** (1)	**3.91** (1)	**2.75** (1)	24.64 (2)	17.66 (2)	39.37 (2)	16.08 (2)	1.57

Matching and Vector Extrema. In $|tokens| \geq 10$, our models achieved better performance than the baselines in all the embedding metrics. In particular, FCRG-BL model performed best, and then FCRG-DT model was the runner up. To sum up, FCRG-DT and FCRG-BL outperformed the baselines in Embedding metrics.

Qualitative Evaluation Next, we conducted another experiment to compare our FCRG-DT with baselines qualitatively. In the experiment, we chose FCRG-DT instead of FCRG-BL since it does not require any additional parameters and had comparable performance with FCRG-BL. We also used $\tau = 10$ to generate responses with at least 10 tokens in all models since lengthy responses are more interesting and informative despite a harder problem.

Human Evaluation Similar to [25], we randomly selected 50 original tweets from the test set. Given each of the original tweets, each of FCRG-DT, SeqAttB and HRED generated 15 responses. Then, one response with the highest probability per model was selected. We chose a pairwise comparison instead of listwise comparison to make easy for human evaluators to decide which one is better. Therefore, we created 100 triplets (original tweet, response$_1$, response$_2$) where one response was generated from our FCRG-DT and the other one was from a baseline. We employed three crowd-evaluators to evaluate each triplet where each response's model name was hidden to the evaluators. Given each triplet, the evaluators independently chose one of the following options: (i) win (response$_1$ is better), (ii) loss (response$_2$ is better), and (iii) tie (equally good or bad). Before labeling, they were trained with a few examples to comprehend the following criteria: (1) the response should fact-check information in the original tweet, (2) it should be human-readable and be free of any fluency or grammatical errors, (3) the response may depend on a specific case or may be general but do not contradict the first two criteria. The majority voting approach was employed to judge which response is better. If annotators rated a triplet with three different answers, we viewed the triplet as a tie. Table 7 shows human evaluation results. The Kappa values show moderate agreement among the evaluators. We conclude that FCRG-DT outperforms SeqAttB and HRED qualitatively.

Case Studies Table 8 presents examples of original tweets, ground-truth D-tweets, and generated responses of the three models. Our FCRG-DT generated more relevant responses with clear fact-checking intention. For example, in the first example, FCRG-DT captured the *uranium* in the original tweet and generated a relevant response. We observed that SeqAttB usually generated non-relevant content. Responses generated by FCRG-DT were more formal than ones generated by the baselines.

Table 7 The results of human evaluation

Opponent	Win	Loss	Tie	Fleiss Kappa
our FCRG-DT vs. SeqAttB	40%	28%	32%	0.725
our FCRG-DT vs. HRED	40%	36%	24%	0.592

Table 8 Examples of original tweets, ground-truth FC tweets, generated responses of our FCRG-DT and two baselines

Pairs of the original tweet (**OT**) and ground-truth **D-tweet**	Generated responses of our FCRG-DT and two baselines
OT: @user <person> sold <number> % of our uranium to russia . that's your candidate . **D-tweet:** except that she did not . url @user	**SeqAttB**: @user no , <person> did not start the birther movement . url **HRED**: @user here you go , read the whole thing . url **FCRG-DT**: @user allegations about hrc are false . she did not have to give uranium to russia : url
OT: @user i just feel bad for poor franklin (?) , all alone on his side of the table … **D-tweet:** @user interesting . url	**SeqAttB**: @user there is no trucker strike in puerto rico url **HRED**: @user this is a fake . sorry folks . url **FCRG-DT**: @user snopes says this is not true . url url
OT: @user that is why i asked the question so i could have an understanding **D-tweet:** @user do use a troll as your source of info url	**SeqAttB**:@user <person> kissed by former klan member : url **HRED**: @user here you go , read the whole thing . url **FCRG-DT**: @user this is a fake quote . sorry folks . url
OT: show me where <person> actually said that , other than this photoshopped picture from the internet . **D-tweet:** @user fake quote . url	**SeqAttB**: @user no, <person> did not start the birther movement. url **HRED**: @user yeah , no , they are not . url **FCRG-DT**: @user this is a fake news site , right ? url
OT: @user also , you are giving me a different set of statistics than the ones i asked for . **D-tweet:** @user here's what he misquoted to start . url	**SeqAttB**: @user here you go , read the facts : url **HRED**: @user here ya go : url read this . url **FCRG-DT**: @user this is a fake quote from <person> . url

3.4 Discussions

Although our proposed models successfully generated responses with fact-checking intention, and performed better than the baselines, there are a few limitations in our work. Firstly, we assumed guardians freely choose articles that they prefer, and then insert corresponding fact-checking URLs into our generated responses. It means we achieved partial automation in a whole fact-checking process. In our future work, we are interested in even automating the process of selecting an fact-checking article based on content of original tweets in order to fully support guardians and automate the whole process. Perhaps, combining both our recommender system and our text generation framework may help us automate the fact-checking process. Secondly, our framework is based on word-based RNNs, leading to a common issue: rare words are less likely to be generated. A feasible solution is using character-level RNNs [9] so that we do not need to replace rare words with <unk> token. In the future work, we will investigate if character-based RNN models work well on our dataset. Thirdly, we only used pairs of an original tweet and a D-tweet without utilizing other data sources such as previous messages in online dialogues. We also tried to use the content of fact-checking articles, but did not improve performance of our models. We plan to explore other ways to utilize the data sources in the future. Finally, there are many original tweets containing URLs pointing to fake news sources (e.g. breitbart.com) but we did not consider them when generating responses. We leave this for future exploration.

4 Conclusions

In this chapter, we presented novel preventive methods to combat fake news by leveraging online users called guardians. By identifying these guardians and analyzing their behavior in posting fact-checking tweets, we built a novel fact-checking URL recommendation model to personalize fact-checking articles and a response generation framework to help guardians fact-check information faster. In the discussion sections, we described possible extensions of our models to achieve better performance. We believe that our work opens new research directions in fake news intervention.

4.1 Contributions

Portions of this chapter are based on work that appeared in the 2018 and 2019 International ACM SIGIR Conference on Research and Development in Information Retrieval (SIGIR) [28, 29, 32].

References

1. Abel, F., Gao, Q., Houben, G.J., Tao, K.: Analyzing temporal dynamics in twitter profiles for personalized recommendations in the social web. In: Proceedings of the 3rd International Web Science Conference, p. 2. ACM (2011)
2. Abel, F., Gao, Q., Houben, G.J., Tao, K.: Semantic enrichment of twitter posts for user profile construction on the social web. In: Extended Semantic Web Conference, pp. 375–389. Springer (2011)
3. Bahdanau, D., Cho, K., Bengio, Y.: Neural machine translation by jointly learning to align and translate. arXiv preprint arXiv:1409.0473 (2014)
4. Banerjee, S., Lavie, A.: Meteor: an automatic metric for mt evaluation with improved correlation with human judgments. In: Proceedings of the ACL Workshop on Intrinsic and Extrinsic Evaluation Measures for Machine Translation and/or Summarization, pp. 65–72 (2005)
5. Chen, J., Nairn, R., Nelson, L., Bernstein, M., Chi, E.: Short and tweet: experiments on recommending content from information streams. In: Proceedings of the SIGCHI Conference on Human Factors in Computing Systems, pp. 1185–1194. ACM (2010)

6. Cho, K., Van Merriënboer, B., Gulcehre, C., Bahdanau, D., Bougares, F., Schwenk, H., Bengio, Y.: Learning phrase representations using RNN encoder-decoder for statistical machine translation. arXiv preprint arXiv:1406.1078 (2014)
7. Ecker, U.K., Lewandowsky, S., Tang, D.T.: Explicit warnings reduce but do not eliminate the continued influence of misinformation. Mem. Cognit. **38**(8), 1087–1100 (2010)
8. Friggeri, A., Adamic, L.A., Eckles, D., Cheng, J.: Rumor cascades. In: ICWSM (2014)
9. Kim, Y., Jernite, Y., Sontag, D., Rush, A.M.: Character-aware neural language models. Association for the Advancement of Artificial Intelligence (AAAI) (2016)
10. Koren, Y., Bell, R., Volinsky, C.: Matrix factorization techniques for recommender systems. Computer **42**(8), 30–37 (2009)
11. Lab, R.: Fact-checking triples over four years. https://reporterslab.org/fact-checking-triples-over-four-years/ (2018)
12. Le, Q., Mikolov, T.: Distributed representations of sentences and documents. In: Proceedings of the 31st International Conference on Machine Learning (ICML-14), pp. 1188–1196 (2014)
13. Lee, K., Mahmud, J., Chen, J., Zhou, M., Nichols, J.: Who will retweet this? Automatically identifying and engaging strangers on twitter to spread information. In: Proceedings of the 19th international conference on Intelligent User Interfaces, pp. 247–256. ACM (2014)
14. Levy, O., Goldberg, Y.: Neural word embedding as implicit matrix factorization. In: Advances in Neural Information Processing Systems, pp. 2177–2185 (2014)
15. Liang, D., Altosaar, J., Charlin, L., Blei, D.M.: Factorization meets the item embedding: regularizing matrix factorization with item co-occurrence. In: Proceedings of the 10th ACM Conference on Recommender Systems, pp. 59–66. ACM (2016)
16. Lin, C.Y.: Rouge: A package for automatic evaluation of summaries. Text Summarization Branches Out (2004)
17. Liu, C.W., Lowe, R., Serban, I., Noseworthy, M., Charlin, L., Pineau, J.: How not to evaluate your dialogue system: An empirical study of unsupervised evaluation metrics for dialogue response generation. In: Proceedings of the 2016 Conference on Empirical Methods in Natural Language Processing, pp. 2122–2132 (2016)
18. Luong, M.T., Pham, H., Manning, C.D.: Effective approaches to attention-based neural machine translation. arXiv preprint arXiv:1508.04025 (2015)
19. Maddock, J., Starbird, K., Al-Hassani, H.J., Sandoval, D.E., Orand, M., Mason, R.M.: Characterizing online rumoring behavior using multi-dimensional signatures. In: Proceedings of the 18th ACM Conference on Computer Supported Cooperative Work & Social Computing, pp. 228–241. ACM (2015)
20. Mikolov, T., Sutskever, I., Chen, K., Corrado, G.S., Dean, J.: Distributed representations of words and phrases and their compositionality. In: Advances in Neural Information Processing Systems, pp. 3111–3119 (2013)
21. Nyhan, B., Reifler, J.: When corrections fail: the persistence of political misperceptions. Polit. Behav. **32**(2), 303–330 (2010)
22. Papineni, K., Roukos, S., Ward, T., Zhu, W.J.: Bleu: a method for automatic evaluation of machine translation. In: Proceedings of the 40th Annual Meeting on Association for Computational Linguistics, pp. 311–318. Association for Computational Linguistics (2002)
23. Rendle, S., Freudenthaler, C., Gantner, Z., Schmidt-Thieme, L.: Bpr: Bayesian personalized ranking from implicit feedback. In: Proceedings of the Twenty-Fifth Conference on Uncertainty in Artificial Intelligence, pp. 452–461. AUAI Press (2009)
24. Serban, I.V., Sordoni, A., Bengio, Y., Courville, A., Pineau, J.: Building end-to-end dialogue systems using generative hierarchical neural network models. arXiv preprint arXiv:1507.04808 (2015)
25. Shang, L., Lu, Z., Li, H.: Neural responding machine for short-text conversation. arXiv preprint arXiv:1503.02364 (2015)
26. Shao, C., Ciampaglia, G.L., Flammini, A., Menczer, F.: Hoaxy: a platform for tracking online misinformation. In: Proceedings of the 25th International Conference Companion on World Wide Web, pp. 745–750. International World Wide Web Conferences Steering Committee (2016)

27. Starbird, K., Palen, L.: Pass it on? Retweeting in mass emergency. In: Proceedings of the 7th International ISCRAM Conference (2010)
28. Vo, N., Lee, K.: The rise of guardians: fact-checking url recommendation to combat fake news. In: Proceedings of the 41st International ACM SIGIR Conference on Research and Development in Information Retrieval (2018)
29. Vo, N., Lee, K.: Learning from fact-checkers: analysis and generation of fact-checking language. In: Proceedings of the 42nd International ACM SIGIR Conference on Research and Development in Information Retrieval, pp. 335–344. ACM (2019)
30. Wang, C., Blei, D.M.: Collaborative topic modeling for recommending scientific articles. In: Proceedings of the 17th ACM SIGKDD International Conference on Knowledge Discovery and Data Mining, pp. 448–456. ACM (2011)
31. Yang, C., Harkreader, R., Zhang, J., Shin, S., Gu, G.: Analyzing Spammers' social networks for fun and profit: a case study of cyber criminal ecosystem on twitter. In: WWW (2012)
32. You, D., Vo, N., Lee, K., Liu, Q.: Attributed multi-relational attention network for fact-checking url recommendation. In: Proceedings of the 28th ACM International Conference on Information and Knowledge Management, pp. 1471–1480. ACM (2019)
33. Zhao, Z., Resnick, P., Mei, Q.: Enquiring minds: Early detection of rumors in social media from enquiry posts. In: Proceedings of the 24th International Conference on World Wide Web, pp. 1395–1405. ACM (2015)

Part III
Trending Issues

Developing a Model to Measure Fake News Detection Literacy of Social Media Users

Julian Bühler, Matthias Murawski, Mahdieh Darvish, and Markus Bick

Abstract Triggered by popular cases such as political election campaigns in the United States of America and the United Kingdom, research on fake news, particularly in the context of social media, has gained growing importance recently. Our chapter deals with the individual user's perspective and places the focus on the competency to detect fake news – the so-called fake news detection literacy. One main challenge in this field is the empirical measurement of such an individual fake news detection literacy. Based on our previous research, we suggest an extended version of a general social media information literacy (SMIL) model which is enriched with respect to the context of fake news, i.e., mainly the evaluation of information. The extended model is empirically tested by applying correlation analyses based on a sample of $n = 96$. The updated construct provides a way to measure fake news detection literacy and offers various avenues for further research that are discussed at the end of the chapter.

Keywords Amazon Mechanical Turk · Correlation analysis · Fake news · Fake news detection · Measurement · Social media · Social media information literacy (SMIL) · User-generated content

1 Introduction

Social media services have become a major source for news [e.g., 1]. Compared to traditional and mostly unidirectional media services (such as printed newspapers or television), these services change the characteristic of distributed information towards being dynamic. Particularly the concept of user-generated content (UGC)

J. Bühler (✉) · M. Murawski · M. Darvish · M. Bick
ESCP Europe Business School Berlin, Berlin, Germany
e-mail: jbuehler@escpeurope.eu; mmurawski@escpeurope.eu; mdarvish@escpeurope.eu; mbick@escpeurope.eu

© Springer Nature Switzerland AG 2020
K. Shu et al. (eds.), *Disinformation, Misinformation, and Fake News in Social Media*, Lecture Notes in Social Networks,
https://doi.org/10.1007/978-3-030-42699-6_11

implies that users can easily modify information, thus allowing them to add their own opinions or even change the meaning dynamically [e.g., 2, 3].

Consequently, one major disadvantage of social media services and the related UGC is that no trusted authority exists which verifies the quality of information distributed through the services' networks. For example, it is relatively easy to produce misleading or false information, which is often referred to as fake news [1]. Safieddine et al. (p. 126) [4] express their concerns in this context by highlighting that the idea of online freedom of expression seems to fail: *"It has allowed totally unprofessional content; developers bombard predominantly passive web content consumers with news, facts, and stories that cannot be easily challenged."* Placing the focus on social media users, Safieddine et al. observe that they *"gradually filter pages, news agencies, or even friends whom they disagree with their political, theological, and/or ethical predispositions"* (p. 126), which could lead, amongst others, to a growing number of parallel realities.

Fake news is omnipresent in today's world and have the potential to cause massive social and monetary damage on every level, i.e., from an individual to a political or societal level [5]. Social media services have become an important instrument during election campaigns since the US election in 2008 [6], and their impact is rising steadily. Two prominent examples of recent political votes, which are discussed, are the Brexit referendum and the election of the president of the United States in 2016. In both cases, fake news was used to manipulate the voters, sometimes even combined with, for example, analyses of social media user profiles [7].

An early fake news detection limits the spread and contributes to trustworthiness of the news ecosystem [8]. However, fake news in social media could mix true with false evidence to support nonfactual claims [9] and create different degrees of fakeness such as half-true, false, etc. [10], increasing the complexity of fake news detection in social media.

Recently, scholars from different disciplines have suggested potential solutions to fight fake news and corresponding damages. From a technical perspective, one promising example is the "right-click authentication" [11–13], which allows the reader to easily check with few clicks the source and reliability of pictures posted online. Other scholars see the social media service providers in charge of ensuring true news [e.g., 4]. Complementing these approaches, we place the competency of the individual social media users in the center of this chapter and propose the concept of *fake news detection literacy*. This is similar to current and ongoing discussions about "media literacy" [14, 15] or "news literacy" [16]. While existing work is mainly of conceptual nature, we offer a concrete way to measure fake news detection literacy.

Our measurement model is based on previous work we conducted to develop a construct to measure the general social media information literacy (SMIL) of a single social media user [17]. In this chapter, we apply and expand the SMIL model according to the context of fake news. We will outline in Sect. 4 that the necessary expansion is mainly relevant for the SMIL sub-category "evaluation". To sum up, the research question (RQ) of this study is.

RQ: How can fake news detection literacy be measured?

The structure of this chapter is as follows. First, the concept of fake news and its current state of research are briefly outlined. After that, our general SMIL construct is presented. This serves as the basis for the following section in which the SMIL construct is extended and empirically tested with three new fake news-related sub-items. The paper ends with a discussion of implications for research and practice and potential applications of the fake news detection literacy model.

2 Fake News

When it comes to the consumption of news, social media services have outpaced traditional sources such as paper-based newspapers or television formats [18]. While social media services on the one hand can offer a more convenient and tailored customer experience, they on the other hand build the basis for fake news. For example, news feeds in social media typically contain public as well as private postings and are intertwined with the online activities of the consumer [19]. This, amongst others, makes it very difficult for the consumer to evaluate the quality of the news.

While fake news as a term is widely adopted, its academic definition is subject to intense discussions [20]. Starting with a very generic definition, Allcot and Gentzkow (p. 213) [1] describe fake news as

news articles that are intentionally and verifiably false, and could mislead readers.

Gelfert [20] employs a broad and deep discussion of the term fake news. He identifies several similarities among extant definitions. First, the medium internet, particularly social media, plays an important role for both creation and dissemination of fake news. Second, fake news do not have any factual basis. Third, fake news are intentionally misleading. He compares and criticizes extant definitions (which is not further considered in this chapter) and suggests his own one:

Fake news is the deliberate presentation of (typically) false or misleading claims as news, where the claims are misleading by design. (p. 108)

The main formats of fake news are images, videos, and text [4]. Furthermore, scholars have elaborated that there are some general characteristics of fake news such as the content, user response, source, and spreaders [21]. Different types of fake news and fake news related terms (gossip, rumors, satire, etc.) show various forms of these characteristics [20–22]. More details about the key characteristics as well as linguistic analyses and user engagement studies of fake news´ properties will be discussed in detail in Sect. 4.1.

Considering the supply side, there seem to be two main motivations for producing fake news [1, 20]: The first one is financial reward. It is possible to draw substantial advertising revenues from clicks on the respective site. A popular case is the one of teenagers in Macedonia who earned thousands of dollars with produced fake news about both Clinton and Trump during their election campaigns [1]. The second motivation is ideological. Taking again the example of political elections, producers of fake news try to advance the politician they favor.

We believe that technical solutions to fight fake news, such as the "right-click authentication" [11–13], are a step into the right direction, but even more important are literate social media users. Literate, or competent, social media users are more likely to detect fake news. This means they are more critical regarding the reliability of news, and they are willing to spend time on conducting required proofs. We call this literacy "fake news detection literacy". It is strongly linked to social media information literacy (SMIL) and, thereby, to the general topic of digital competencies [23]. However, in order to answer further research questions, for example, about the relation of individual fake news detection literacy and actual fake news detection, a measurement model of individual fake news detection literacy is required.

In a previous paper [17], we have developed a measurement model of SMIL, which is briefly summarized and explained in the next section. Taking this as the basis, we will then suggest an expanded measurement model that is applied to the context of fake news detection.

3   Introduction of SMIL

3.1   *Development of SMIL Definition*

In our previously mentioned research article, we developed a new construct from scratch, which is meant to measure a user's information literacy regarding social media (SMIL). The article is theoretically motivated and based on MacKenzie et al. [24] who provide a detailed step-by-step guideline for construct development in general. Our main argument for choosing this approach for the core SMIL construct was the possibility of starting at an initial stage for such a core construct, rather than relying on an existing construct that is expanded, but not completely suitable. SMIL itself in its current state is a suitable basis for expansions such as fake news, though, because it regards the specific elements social media bring along, especially its dynamic processes [18, 25].

MacKenzie et al. [24] describe the construct development process in several steps, which we applied for SMIL consecutively. Based on the guidelines of Webster and Watson [26], we conducted an extensive review of existing literature in the field of social media research. Scanning various scientific databases like Scopus, ScienceDirect, JSTOR, EBSCO and others was a key element due to the fast-changing environment, with new social media services being on the rise and providing new functions to its users [18]. Regarding particularly the perspective of literacy in the context of social media, we applied the following search query:

```
information literacy AND social media OR construct*OR measure*
```

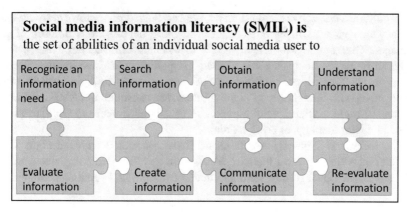

Fig. 1 Definition of social media information literacy (SMIL) [17]

Starting with in total 88 articles which cover a rather wide range of academic fields and aspects of literacy (i.e., *metaliteracy*, *transliteracy* [27], or *reading literacy* [28]), we extracted relevant keywords in multiple iterations. We then clustered these keywords according to the description of MacKenzie et al. [24] to create a holistic definition of SMIL (Fig. 1).

3.2   SMIL Item Development, Evaluation and Refinement

Based on the SMIL definition, the next couple of steps in MacKenzie et al.'s [24] guideline recommend the derivation of individual items. This includes their phrasing as well as their initial testing and refinement. By literature screening, we extracted a portfolio of existing items from academic sources which we then completed with new items derived from our SMIL definition. This led to a total of 40 unique items that were each associated with one of the clusters of abilities that form our SMIL definition.

The item evaluation was then applied following the quantitative approach of Hinkin and Tracey [29]. We calculated the items' content validity with the help of results from 59 surveys that we conducted. This outnumbers the threshold of 50 which the authors suggest. With the method of one-way repeated measures ANOVAs, we tested how statistically significant the items within the cluster are. Consequently, we rephrased 14 items according to Wieland et al. [30] to increase content validity.

Whereas our complete step-by-step procedure following MacKenzie et al. [24] is explained in detail in our original SMIL article [17], we briefly summarize the results of the final refinement step, which ultimately formed the core SMIL construct. We enlarged our empirical setting to conduct reliability checks by spreading the survey to a different target group from a different country to include cross-cultural perspectives. Additionally, we capitalized on a crowdsourcing campaign which gave us insights from an even more heterogenous number of participants. The sample was large enough to evaluate our 40-item scale with an exploratory factor analysis (EFA) based on participants' item ratings. Results based on the Eigenvalues revealed an optimum of seven to eight clusters which corresponds with the number of eight abilities that constitute our SMIL definition. Within the eight clusters, we eliminated in a final refinement process those items with low factor loadings or cross-loadings.

The following Table 1 presents all 40 items associated with the eight SMIL clusters, of which 19 form the core SMIL item set. The remaining items shaded with a dark grey background were eliminated from the first version of the item set

Table 1 Overview of 40 SMIL items

Code	Phrase
REC_1	I am able to recognize the information I need.
REC_2	I am able to realize my need for information.
REC_3	I am able to recognize the information I do not need.
SEA_1	I am able to decide where and how to find the information I need.
SEA_2	I am able to technically access information.
SEA_3	I am able to apply appropriate search strategies (e.g., use of meaningful keywords).
SEA_4	I am able to limit search strategies (e.g., date, hashtag, user).
SEA_5	I am able to choose appropriate sources when searching for information.
OBT_1	I am able to collect information.
OBT_2	I am able to retrieve information.
OBT_3	I am able to choose appropriate information.
UND_1	I am able to interpret information.
UND_2	I am able to find consensus among sources.
UND_3	I am able to understand the intention of information.
UND_4	I am able to identify points of agreement and disagreement among information sources.
UND_5	I am able to understand type and delivery mode of information.
EVAL_1	I am able to evaluate the relevance of information.
EVAL_2	I am able to evaluate the credibility of information.
EVAL_3	I am able to evaluate the accuracy of information.
EVAL_4	I am able to evaluate the quality of information.
EVAL_5	I am able to identify if information is a fake.
EVAL_6	I am able to identify if information is a rumor.
CREAT_1	I am able to rephrase information to clarify its meaning.
CREAT_2	I am able to create context for information.
CREAT_3	I am able to modify identified information.
CREAT_4	I am able to merge information.
CREAT_5	I am able to change the scope by reducing information.

(continued)

Table 1 (continued)

Code	Phrase
CREAT_6	I am able to enrich identified information.
CREAT_7	I am able to design information.
COMM_1	I am able to display information for a given audience.
COMM_2	I am able to share information with others.
COMM_3	I am able to provide feedback.
COMM_4	I am able to communicate information safely and securely.
COMM_5	I am able to exchange information.
COMM_6	I am able to provide constructive criticism to other users.
REVAL_1	I am able to use reflective practices in order to re-evaluate information.
REVAL_2	I am able to evaluate users' reaction on my content.
REVAL_3	I am able to evaluate information from interaction with other users.
REVAL_4	I am able to reconsider my existing evaluation of information.
REVAL_5	I am able to identify the benefits of re-evaluating information.

Adapted from Murawski et al. [17]

and, thus, the entire initial version of the SMIL construct after the aforementioned validity tests.

Especially items of the clusters *obtain information* and *understand information* were deleted due to the factor loadings. For *evaluate information*, we could identify two separate factors accounting for the cluster of which one addresses rather abstract evaluation and one refers to concrete action of evaluation accomplished by users. Beside the content-related motivation to expand the core SMIL construct towards fake news, this split of evaluation clusters reinforces the decision also from an empirical perspective. Because within *evaluate information*, the items *Eval_5* and *Eval_6* are directly linked to the evaluation of faked information and rumors. Thus, we proceed in the next section with the application of the core SMIL construct to the context of fake news by primarily expanding *EVAL_5*.

4 Applying SMIL to the Context of Fake News Detection

4.1 Key Characteristics of Fake News

There are three generally agreed upon characteristics of fake news: its content style, the user engagement with it, and the source users publishing it [21].

The semantic characteristics of fake news content vary across different types of fake news and fake news related terms such as gossip, rumors, hoaxes, satire and etc. [20–22]. However, scholars focusing on text analysis of fake news have found some linguistic cues regarding pronouns, conjunctions and word patterns [21]. Fake news contains personal pronouns and words associated with negative emotions such as swearwords [31]. High uncertainty or many typographical errors are other cues for news content of low quality [32]. In addition, low quality and high informality of the headline are two characteristics of fake news [33]. For instance, in a high quality news article there is a similarity between the headline and the body-text [22]. However, swear words ('damn'), net speaks ('btw' and' 'lol'), assents ('OK'), non fluencies ('er', 'hm', and, 'umm'), and fillers ('I mean' and 'you know') are signs of informality in the headline [32].

The second characteristic is the emotional response that news generates. Fake news contains opinionated and emotionally provoking language generating a sense of confusion [21]. Furthermore, sensational or even faked visual impressions (e.g. images and videos) can be employed to provoke specific emotional responses from consumers [34].

The third characteristic of fake news is the source promoting it. Starting from the URL structure, the source media and the author of the news, different properties of a publishing source should be examined [21]. Corresponding with the definition of fake news suggested by Gelfert, sources and channels promoting fake news are misleading "by design", to manipulate the audience's cognitive process [20]. For example news posted on an unreliable platform and promoted by unreliable users is more likely to be fake news than news published by authoritative and credible spreaders [32].

Table 2 Fake news expansion of the core SMIL construct

SMIL core cluster: *Evaluate*	
Item Code	**Item Phrase**
EVAL_1	I am able to evaluate the relevance of information.
EVAL_2	I am able to evaluate the credibility of information.
EVAL_3	I am able to evaluate the accuracy of information.
EVAL_4	I am able to evaluate the quality of information.
EVAL_5	**I am able to identify if information is a fake.**
EVAL_6	I am able to identify if information is a rumor.

SMIL cluster expansion: *Fake News*	
Sub-item Code	**Sub-item Phrase**
EVAL_5 – SUB 1	I am able to identify differences between headline and text-body of news. [22, 35]
EVAL_5 – SUB 2	I am able to distinguish satire and fake news. [21, 22]
EVAL_5 – SUB 3	I am able to identify automated accounts (bot) spreading information.[36]

4.2   *Expansion of the SMIL Core Construct*

The described characteristics of fake news ultimately lead to the expansion of the core SMIL construct. Expanding the construct is advisable because of the dynamic environment fake news predominantly appears in, which is social media as previously characterized. The specific expansion is visualized in Table 2, with modifications precisely originating from the closest item of the core construct, *EVAL_5*.

The expansion of *EVAL_5* is the major change we propose regarding the application of our SMIL construct towards fake news detection. This core item of SMIL is directly linked to the topic and we differentiate between three new aspects that shed more light on the evaluation ability of a social media user regarding fake news. We call these aspects sub-items, i.e., *EVAL_5 – SUB_1* to *EVAL_5 – SUB_3*, which are meant to represent a hierarchical graduation between the existing superordinate core item and the new ones dedicated explicitly to areas of fake news.

Whereas the headlines itself can sometimes help revealing fake news based on quality and formality on their own [33], the interplay between the headline and the main text-body is an even stronger criterion for evaluation. Especially the mix between a headline that is intended to raise awareness by capitalizing on clickbait elements and a text-body that refers to rather accurate content is important [35]. Consequently, *EVAL_5 – SUB_1* addresses this ability. The second sub-item *EVAL_5 – SUB 2* refers to a different aspect, the ability to distinguish between truly fake news and satirical content with partly similar patterns, but also recognizable differences [21], e.g., the motivation for spreading the news. The third new sub-item *EVAL_5 – SUB 3* represents the ability of a social media user to identify whether the source of an information is a real person or only an automated system. Bot networks are the most common thread in terms of fake news, which is specifically regarded with this sub-item [36].

Besides evaluation, other clusters of this core construct are also affected. Particularly those dealing with the quality of information sources, i.e., items SEA_5 (appropriate choice of information sources), UND_2 and UND_4 (interplay between multiply sources). Additionally, the remaining items of *re-evaluation* can be of relevance in terms of fake news identification. The exchange with and reactions from other users can indicate a previously not recognized fake news appearance. We do not see the need to state particular sub-items for these core items. But while applying the SMIL construct in the context of fake news, including the new sub-items within the evaluation cluster, special attention should be paid to these items of the core construct as well.

4.3   Empirical Evaluation of the Extended Model

In a next step, we empirically tested our model extension towards fake news. For this, we used Amazon Mechanical Turk[1] (MTurk) to collect data. MTurk has gained popularity among researchers [37], as it enables both to conveniently access a large pool of potential respondents, and to receive responses without any time delays at comparably low costs (particularly compared to paper-based study designs). However, aside from these advantages, it must be noted that the quality of the responses is a critical aspect. Thus, as suggested in related literature [38], we integrated two direct quality checks in our survey. First, we implemented a captcha questions which had to be answered before the survey begins. This question was designed as a simple calculation, e.g., $37 + 3 =$ ___, that nevertheless requires human knowledge, and therefore does not allow for instance bots entering the survey. Second, we implemented a test question in our total set of question which reads as follows: *I am able to develop the next Facebook. If you are reading this question, select I completely disagree.* Answers different from *I completely disagree* on this question led to an exclusion of the respondent from the dataset, as it can be assumed that questions were not read.

In addition to these two direct checks, we considered the time to fill the questionnaire by a respondent ('input time') as an indirect check. Based on test runs, we defined 2:30 min as the minimum input time that is required to enter meaningful answers. For questionnaires with less than 2:30 min input time, a high chance for random clicking by the respondent is given. We therefore eliminated these respondents from our dataset.

The initial dataset for the empirical analysis consisted of $n = 172$ completed responses at the time of survey closure. Based on the aforementioned exclusion criteria, we reduced the sample to $n = 96$ responses because 59 participants did not reach the time threshold, additional 16 did not respond to the test question appropriately and one respondent claimed to be 6 years old, thus underage. A majority of the 96 respondents stated that they are from the USA or India, approximately 58% stated that they are male (39% female, 3% did not disclose a gender). The average response time for the survey including the specific fake news extension of the research model was 4:45 min (sd = 2:08).

All 40 core items and the three additional fake news items were measured on a 5-point Likert scale ranging from 1 (*I completely disagree*) to 5 (*I completely agree*). Thus, we can assume equidistance and calculate the average mean for the three fake news items. The highest score of 4.52/5.00 was measured for EVAL_5– SUB 1, revealing a strong confidence within the sample of being able to differentiate between a news headline and its text-body in contrast. Results for the remaining two items also indicate rather strong confidence in the respective abilities, but are lower for EVAL_5– SUB 2 (4.16/5.00, "ability to distinguish between satire

---

[1] Accessible at https://www.mturk.com

Table 3 Correlation analysis between the three fake news-related sub-items

Correlation analysis (Pearson)		
	EVAL_5 – SUB 2	EVAL_5 – SUB 3
EVAL_5 – SUB 1	.356**	−.0,13 n.s.
EVAL_5 – SUB 2	–	.239*

EVAL_5 – SUB 1: I am able to identify differences between headline and text-body of news [21, 34]
EVAL_5 – SUB 2: I am able to distinguish satire and fake news [20, 21]
EVAL_5 – SUB 3: I am able to identify automated accounts (bot) spreading information [36]
n.s. not significant
*Correlation is significant at the 0.05 level
**Correlation is significant at the 0.01 level

and fake news") and EVAL_5– SUB 3 (3.61/5.00, "ability to identify automated accounts/bots").

Our results indicate both similarities between the three fake news items, as all have above-average means, but also differences because these means still tiered. Thus, we proceeded with a correlation analysis using IBM SPSS Statistics 25 to investigate the relation between the new model items. We calculated bivariate Pearson correlation coefficients and tested for two-tailed significance (Table 3).

Our results demonstrate a strong positive correlation that is highly significant between EVAL_5 – SUB 1 and EVAL_5 – SUB 2 in the sample. It suggests that social media users with the ability to identify differences between a news headline and its text are also able to differentiate between satirical news elements and those that are faked. However, these users not necessarily have the ability to recognize automated news created by bots as there is no significant relation between these sub-items. Although weaker, a statistically significant correlation exists as well between EVAL_5 – SUB 2 and EVAL_5 – SUB 3. This indicates that social media users with a higher ability to distinguish between satire and fake news likely are able to identify bots.

In summary, our empirical correlation analysis' results show that the three fake news-related sub-items of the core item EVAL_5 are significantly linked with each other. This supports our initial claim that the core SMIL model should be extended if the fake news phenomenon is to be analyzed more precisely. We suggested three new sub-items which are statistically related with each other and which might serve as standard items for future empirical fake news studies based on SMIL.

5 Contributions, Limitations, and Future Work

The suggested model to measure fake news detection literacy is of value for both researchers and practitioners. Considering research, we contribute to the field of the digitization of the individual and corresponding micro-foundations, as we do not investigate a digitalization phenomenon on the organizational but the individual level [39]. Our model establishes a link to previous theoretical work about rigor construct development that originates from scratch [24], it is based on the already initially tested core SMIL construct [17], and it allows various other applications linked to fake news or related topics within the social media environment. The initially raised research question can be answered with the newly introduced sub-items, which are derived from theory and turned out to be meaningful considering our empirical testing (see Sect. 4). Thus, fake news detection literacy can be measured by adding the three sub-items to the original core SMIL construct and by specifically regarding the discussed core items of the *search*, *understand*, and *re-evaluate* cluster.

Practitioners however can benefit from a concrete set of items, which allows them a hands-on approach towards news classification. We provide them with our expanded SMIL construct, an instrument that regards the dynamic changes in the fast-paced social media context. One example that demonstrates the value of the expanded item set would be the concepts of competency and literacy, which are of growing importance in the educational system. Curricula are often designed to impart competencies but not pure content, which also corresponds to the requirements formulated by accreditation agencies. Thus, when curriculum designers and teachers could assess the fake news detection literacy of pupils or students, the development of respective courses and materials could be more focused and tailored. This also applies to other (commercial) training providers, covering different age groups and subjects. Human resource managers is another stakeholder group for our model. Similar to educational institutions, companies might also be interested in the fake news detection literacy of their employees in the age of digital information. Information has become a critical resource for many businesses, which underscores the importance if literate employees in this regard.

Considering the limitations of our study, we are aware of the fact that our empirical assessment of the three added sub-items (see Sect. 4.3) should be interpreted as a first step towards validation. In upcoming studies, the newly derived sub-items could either replace the EVAL_5 core item entirely or could be added to the core item set again to replicate our study design. Going beyond our design, empirical causality could be tested, e.g., with an explorative and confirmatory factor analysis, as it has already been performed for the core model by [17].

Another more general limitation of our study is the assumption that self-assessment is a suitable approach to measure literacy or competency. We are aware that self-assessment is always at risk of bias, and therefore we vote for combining it with alternative approaches such as experiments or observations. However, our set of items could also serve as the basis for other approaches such as interviews (e.g., our items could be used to develop an interview guideline).

Aside from this general view on future research opportunities, we have identified two more specific application areas for our fake news detection literacy model. First, the postulated positive relationship between fake news detection literacy and actual fake news detection performance should be investigated. On the one hand, we believe that it is impossible to identify every piece of fake news, on the other hand we believe that a social media user with a certain level of fake news detection literacy should be able to identify most fake news. A corresponding research question could be *What is the necessary level of fake news detection literacy?* Following this line of argumentation, it may not be useful to aim for the highest possible level of fake news detection literacy. Instead, it could be more important to focus on dynamic training approaches, which brings us to the second specific avenue for further research. The overarching question here would be *How can we impart fake news detection literacy?* This question is not trivial to answer, given the extremely dynamic and innovative field of fake news production, which often uses text mining and other big data analytics to provide the consumer with the "right" fake news. Another related question is the one of responsibility. *Who is in charge of imparting fake news detection literacy, or, on a general level, social media information literacy?* Is it the teachers, who often lack these literacies themselves? Or is it the parents? Or is it the employers? Or is every individual user responsible for his- or herself? Given this complex setting, we believe that this topic requires interdisciplinary research efforts particularly from the fields of information systems, psychology, and education.

In the age of digitalization and information, fake news can cause massive damages to an individual person, to a company, or to an entire society. Empowering people with the necessary competencies, more precise with fake news detection literacy, is therefore a key challenge and we believe that our measurement model marks a valuable contribution towards the next level of understanding.

References

1. Allcott, H., Gentzkow, M.: Social media and fake news in the 2016 election. J. Econ. Perspect. **31**, 211–236 (2017)
2. Baur, A.W., Lipenkova, J., Bühler, J., Bick, M.: A novel design science approach for integrating Chinese user-generated content in non-Chinese market intelligence. In: Proceedings of 36th International Conference on Information Systems (ICIS). Fort Worth, TX, USA (2015)
3. Mikalef, P., Sharma, K., Pappas, I.O., Giannakos, M.N.: Online reviews or marketer information? An eye-tracking study on social commerce consumers. In: Kar, A.K., Ilavarasan, P.V., Gupta, M.P., Dwivedi, Y.K., Mäntymäki, M., Janssen, M., Simintiras, A., Al-Sharhan, S. (eds.) Digital Nations – Smart Cities, Innovation, and Sustainability, vol. 10595, pp. 388–399. Springer International Publishing, Cham (2017)
4. Safieddine, F., Masri, W., Pourghomi, P.: Corporate responsibility in combating online misinformation. Int. J. Adv. Comput. Sci. Appl. **7**, 126–132 (2016)
5. Lazer, D.M.J., Baum, M.A., Benkler, Y., Berinsky, A.J., Greenhill, K.M., Menczer, F., Metzger, M.J., Nyhan, B., Pennycook, G., Rothschild, D., et al.: The science of fake news. Science. **359**, 1094–1096 (2018)
6. Bühler, J., Bick, M.: The impact of social media appearances during election campaigns. In: Proceedings of 19th Americas Conference on Information Systems (AMCIS). Chicago, IL, USA (2013)
7. Blesik, T., Murawski, M., Vurucu, M., Bick, M.: Applying big data analytics to psychometric micro-targeting. In: Bhattacharyya, S., Bhaumik, H., Mukherjee, A., De, S. (eds.) Machine Learning for Big Data Analysis, pp. 1–30. De Gruyter, Berlin, Boston (2018)

8. Shu, K., Wang, S., Liu, H.: Beyond news contents: the role of social context for fake news detection. In: Proceedings of the Twelfth ACM International Conference on Web Search and Data Mining, pp. 312–320. ACM (2019)
9. Shu, K., Sliva, A., Wang, S., Tang, J., Liu, H.: Fake news detection on social media: a data mining perspective. ACM SIGKDD Explorations Newsl. **19**, 22–36 (2017)
10. Karimi, H., Roy, P., Saba-Sadiya, S., Tang, J.: Multi-source multi-class fake news detection. In: Proceedings of the 27th International Conference on Computational Linguistics, pp. 1546–1557 (2018)
11. Pourghomi, P., Halimeh, A.A., Safieddine, F., Masri, W.: Right-click Authenticate adoption: The impact of authenticating social media postings on information quality. In: 2017 International Conference on Information and Digital Technologies (IDT), pp. 327–331. IEEE (2017)
12. Pourghomi, P., Safieddine, F., Masri, W., Dordevic, M.: How to stop spread of misinformation on social media: Facebook plans vs. right-click authenticate approach. In: 2017 International Conference on Engineering & MIS (ICEMIS), pp. 1–8. IEEE (2017)
13. Dordevic, M., Safieddine, F., Masri, W., Pourghomi, P.: Combating misinformation online: identification of variables and proof-of-concept study. In: Dwivedi, Y.K., Mäntymäki, M., Ravishankar, M.N., Janssen, M., Clement, M., Slade, E.L., Rana, N.P., Al-Sharhan, S., Simintiras, A.C. (eds.) Social Media: The Good, the Bad, and the Ugly, vol. 9844, pp. 442–454. Springer International Publishing, Cham (2016)
14. Frechette, J.: Keeping media literacy critical during the post-truth crisis over fake news. Int. J. Crit. Media Literacy. **1**, 51–65 (2019)
15. Mihailidis, P., Viotty, S.: Spreadable spectacle in digital culture: civic expression, fake news, and the role of media literacies in "post-fact" society. Am. Behav. Sci. **61**, 441–454 (2017)
16. Bonnet, J.L., Rosenbaum, J.E.: "Fake news," misinformation, and political bias: Teaching news literacy in the 21st century. Commun. Teach. 34, 103–108 (2020). https://doi.org/10.1080/17404622.2019.1625938
17. Murawski, M., Bühler, J., Böckle, M., Pawlowski, J., Bick, M.: Social media information literacy – what does it mean and how can we measure it? In: Pappas, I.O., Mikalef, P., Dwivedi, Y.K., Jaccheri, L., Krogstie, J., Mäntymäki, M. (eds.) Digital transformation for a sustainable society in the 21st century. 18th IFIP WG 6.11 Conference on e-Business, e-Services, and e-Society, I3E 2019, Trondheim, Norway, September 18–20, 2019, proceedings, vol. 11701, pp. 367–379. Springer, Cham (2019)
18. Bühler, J., Bick, M.: Name it as you like it? Keeping pace with social media something. Electron. Markets. **28**, 509–522 (2018)
19. Bergström, A., Jervelycke Belfrage, M.: News in social media. Digital Journalism. **6**, 583–598 (2018)
20. Gelfert, A.: Fake news: a definition. Informal Logic. **38**, 84–117 (2018)
21. Ruchansky, N., Seo, S., Liu, Y.: CSI: a hybrid deep model for fake news detection. In: Proceedings of the 2017 ACM on Conference on Information and Knowledge Management (CIKM), pp. 797–806. ACM (2017)
22. Potthast, M., Kiesel, J., Reinartz, K., Bevendorff, J., Stein, B.: A stylometric inquiry into hyperpartisan and fake news. In: Proceedings of 56th Annual Meeting of the Association for Computational Linguistics. Melbourne, Australia (2018)
23. Murawski, M., Bick, M.: Digital competences of the workforce – a research topic? Bus. Process Manage. J. **23**, 721–734 (2017)
24. MacKenzie, S.B., Podsakoff, P.M., Podsakoff, N.P.: Construct measurement and validation procedures in MIS and behavioral research. Integrating new and existing techniques. MIS Q. **35**, 293–334 (2011)
25. Baumöl, U., Hollebeek, L., Jung, R.: Dynamics of customer interaction on social media platforms. Electron. Markets. **26**, 199–202 (2016)
26. Webster, J., Watson, R.T.: Analyzing the past to prepare for the future. Writing a literature review. MIS Q. **26**, xiii–xxiii (2002)

27. Brage, C., Lantz, A.: A re-conceptualisation of information literacy in accordance with new social media contexts. In: Callaos, N. (ed.) Proceedings/IMSCI'13, the 7th International Multi-Conference on Society, Cybernetics and Informatics, July 9–12, 2013, Orlando, Florida, USA; [the main conference in the context of the IMSCI 2013 Multi-Conference is the 11th International Conference on Education and Information Systems, Technologies and Applications. EISTA 2013], pp. 217–222. Winter Garden, Fla (2013)
28. Fahser-Herro, D., Steinkuehler, C.: Web 2.0 literacy and secondary teacher education. J. Comput. Teach. Educ. **26**, 55–62 (2010)
29. Hinkin, T.R., Tracey, J.B.: An analysis of variance approach to content validation. Organ. Res. Methods. **2**, 175–186 (1999)
30. Wieland, A., Durach, C.F., Kembro, J., Treiblmaier, H.: Statistical and judgmental criteria for scale purification. Supply Chain Manage. Int. J. **22**, 321–328 (2017)
31. Kar, A.K., Ilavarasan, P.V., Gupta, M.P., Dwivedi, Y.K., Mäntymäki, M., Janssen, M., Siminti-ras, A., Al-Sharhan, S. (eds.): Digital Nations – Smart Cities, Innovation, and Sustainability. Springer International Publishing, Cham (2017)
32. Zhou, X., Zafarani, R.: Fake news: a survey of research, detection methods, and opportunities. arXiv preprint arXiv:1812.00315 (2018)
33. Ainley, J., Fraillon, J., Schulz, W., Gebhardt, E.: Conceptualizing and measuring computer and information literacy in cross-national contexts. Appl. Meas. Educ. **29**, 291–309 (2016)
34. Shu, K., Liu, H.: Detecting fake news on social media. Synth. Lect. Data Min. Knowl. Discovery. **11**, 1–129 (2019)
35. Bourgonje, P., Schneider, J.M., Rehm, G.: From clickbait to fake news detection: an approach based on detecting the stance of headlines to articles. In: Proceedings of the 2017 Workshop: Natural Language Processing meets Journalism, NLPmJ@EMNLP. Copenhagen, Denmark (2017)
36. Edwards, C., Edwards, A., Spence, P.R., Shelton, A.K.: Is that a bot running the social media feed? Testing the differences in perceptions of communication quality for a human agent and a bot agent on Twitter. Comput. Hum. Behav. **33**, 372–376 (2014)
37. Buhrmester, M., Kwang, T., Gosling, S.D.: Amazon's Mechanical Turk: a new source of inexpensive, yet high-quality data? In: Kazdin, A.E. (ed.) Methodological Issues and Strategies in Clinical Research, pp. 133–139. American Psychological Association, Washington, DC (2016)
38. Mason, W., Suri, S.: Conducting behavioral research on Amazon's Mechanical Turk. Behav. Res. Methods. **44**, 1–23 (2012)
39. Matt, C., Trenz, M., Cheung, C.M.K., Turel, O.: The digitization of the individual: conceptual foundations and opportunities for research. Electron. Markets. **29**(3), 315–322 (2019)

BaitWatcher: A Lightweight Web Interface for the Detection of Incongruent News Headlines

Kunwoo Park, Taegyun Kim, Seunghyun Yoon, Meeyoung Cha, and Kyomin Jung

Abstract In digital environments where substantial amounts of information are shared online, news headlines play essential roles in the selection and diffusion of news articles. Some news articles attract audience attention by showing exaggerated or misleading headlines. This study addresses the *headline incongruity* problem, in which a news headline makes claims that are either unrelated or opposite to the contents of the corresponding article. We present *BaitWatcher*, which is a lightweight web interface that guides readers in estimating the likelihood of incongruence in news articles before clicking on the headlines. BaitWatcher utilizes a hierarchical recurrent encoder that efficiently learns complex textual representations of a news headline and its associated body text. For training the model, we construct a million scale dataset of news articles, which we also release for broader research use. Based on the results of a focus group interview, we discuss the importance of developing an interpretable AI agent for the design of a better interface for mitigating the effects of online misinformation.

Keywords Online news · Deep learning · Misinformation · Headline incongruity · Browser extension

K. Park
Qatar Computing Research Institute, Doha, Qatar

T. Kim · M. Cha (✉)
Korea Advanced Institute of Science and Technology, Daejeon, Republic of Korea

Institute for Basic Science, Daejeon, Republic of Korea
e-mail: jgdgj@kaist.ac.kr; mcha@ibs.re.kr

S. Yoon · K. Jung
Seoul National University, Seoul, Republic of Korea
e-mail: mysmilesh@snu.ac.kr; kjung@snu.ac.kr

© Springer Nature Switzerland AG 2020
K. Shu et al. (eds.), *Disinformation, Misinformation, and Fake News in Social Media*, Lecture Notes in Social Networks,
https://doi.org/10.1007/978-3-030-42699-6_12

229

1 Introduction

The dissemination of misleading or false information in news media has become a critical social problem [25]. Because information propagation online lacks verification processes, news contents that are rapidly disseminated online can put veiled threats to society. In digital environments that are under information overload, people are less likely to read or click on the entire contents; instead, they read only news headlines [16]. A substantial amount of news sharing is headline-based, where people circulate news headlines without necessarily having checked the full news story. News headlines are known to play an essential role in making first impressions on readers [32], and these first impressions have been shown to persist even after the full news content has been read [11]. Therefore, if a news headline does not correctly represent the full news story, it could mislead readers into the promotion of overrated or false information, which becomes hard to revoke.

This paper tackles the problem of headline incongruence [7], where a news headline makes claims that are unrelated to or distinct from the story in the corresponding body text. Figure 1 shows such an example, where the catchy news headline promises to describe the benefits of yoga, yet the body text is mainly an advertisement for a new yoga program. While this mismatch can be recognized when people read both the headline and the body text, it is almost impossible to detect it before clicking on the headline in online platforms. Incongruent news headlines make not only incorrect impressions on readers [11] but also

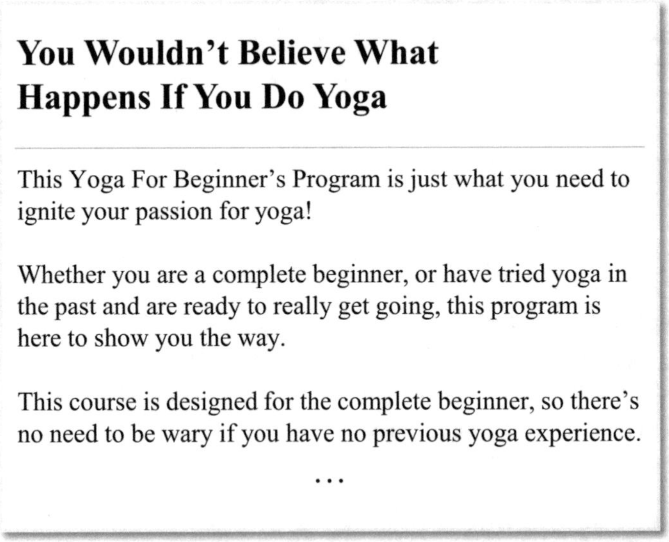

You Wouldn't Believe What Happens If You Do Yoga

This Yoga For Beginner's Program is just what you need to ignite your passion for yoga!

Whether you are a complete beginner, or have tried yoga in the past and are ready to really get going, this program is here to show you the way.

This course is designed for the complete beginner, so there's no need to be wary if you have no previous yoga experience.

· · ·

Fig. 1 An example of news article with the incongruent headline

become problematic when they are shared on social media, where most users share content without reading the full text [16]. Therefore, the development of an automated approach that detects incongruent headlines in news articles is crucial. Identifying incongruent headlines in advance will more effectively assist readers in selecting which news stories to consume and, thus, will decrease the likelihood of encountering unwanted information.

Previous research has tried to detect misleading headlines by either analyzing linguistic features of news headlines [3, 6] or analyzing textual similarities between news headlines and body texts [14, 42]. However, the lack of a large-scale public dataset hinders the development of sophisticated deep learning approaches that will be better suited for such challenging detection tasks, which typically require a million-scale dataset across various domains [19, 27]. This study aims at filling this gap by proposing an automated approach for generating a million-scale dataset for headline incongruity, developing deep learning approaches that are motivated by hierarchical structures of news articles, and evaluating the model in the wild by developing a lightweight web interface that estimates the likelihood of an incongruent news headline.

Our contributions are summarized as follows:

1. We develop a million-scale dataset for the incongruent headline problem, which covers almost all of the news articles that were published in a nation over 2 years. The corpus is composed of pairs of news headlines and body texts along with the annotated incongruity labels. The automatic approach for annotation can be applied to any news articles in any language and, therefore, will facilitate future research on the detection of headline incongruity in a broader research community.
2. We propose deep hierarchical models that encode full news articles from the word-level to the paragraph-level. Experimental results demonstrate that our models outperform baseline approaches that were proposed for similar problems. We extensively evaluate our models with real data. Manual verification successfully demonstrates the efficacy of our dataset in training for incongruent headline detection.
3. To facilitate news reading in the wild, we present BaitWatcher — a lightweight web interface that presents the prediction results that are obtained based on deep learning models *before* readers click on news headlines. Along with this study, implementation details and codes will be shared. BaitWatcher is platform-agnostic; hence, it can be applied to any online news service. The results of a user study of focus group interviews not only support the effectiveness of the BaitWatcher web interface but also reveal a need for the development of interpretable models.

2 Related Work

2.1 Learning-Based Approaches for Detecting Misleading Headlines

Interest has been growing in the automatic detection of misleading headlines. Previous studies constructed datasets that were annotated by journalists or crowd-sourced workers and proposed machine learning approaches. For example, a recent study suggested a co-training method for the detection of ambiguous and misleading headlines from pairs that consist of a title and a body text [43]. We review the literature for each type of misinformation and its relation to the headline incongruence problem.

A series of studies have focused on the detection of *clickbait headlines*, which are a type of web content that attracts an audience and induces them to click on a link to a web page [6]. There is no single and concise definition in the literature; clickbait is regarded as an umbrella term that describes many techniques for attracting attention and invoking curiosity to entice the reader to click on a headline [24]. One study [4] released a manually labeled dataset and developed an SVM model for the prediction of clickbait based on linguistic patterns of news headlines. Using this dataset, other researchers suggested a neural network approach that predicts the clickbait likelihood [35]. A national-level clickbait challenge was held, where the objective was to identify social media posts that entice readers to click on a link [10]. The significant difference between clickbait and headline incongruence is that clickbait is characterized solely by the headline, whereas an incongruent headline defined by the relationship between the headline and the body text. These definitions are not mutually exclusive: a clickbait headline can also be incongruent with its main article. Clickbait headlines may be acceptable if they represent corresponding body texts accurately; however, the consequences can be more severe if catchy headlines mislead people with incorrect information.

The detection of headline incongruence is also related to the stance detection task, which aims at identifying the stance of specified claims against a reference text. The Emergent project [14] provides a dataset of 300 rumored claims and 2,595 associated news articles, each of which is labeled by journalists to indicate whether the stance of the article is *for*, *against*, or *observing* the claim. The Fake News Challenge 2017 was held to promote the development of methods for estimating the stance of a news article [13]. This dataset provides 50,000 pairs of headlines and body texts that were generated from 1,683 original news articles. Each data entry is annotated with one of the following four stances: *agree*, *disagree*, *discuss*, and *unrelated*. While many teams attempted to employ deep learning models (e.g., [8, 34]), the winning model was a simple ensemble approach that combines predictions from XGBoost [5] that are based on hand-designed features and a deep convolution dual encoder that independently learns word representations from headlines and body texts using convolutional neural networks.

Stance detection is technically similar to the headline incongruence problem in that they consider the textual relationship between a headline (claim) and the corresponding body text (reference text). It may be possible to transform three or four stances into binary classes, such as *related* and *unrelated*. But most available datasets cannot be directly utilized for the training of deep learning models for the headline incongruence problem because a *related* headline can also be *incongruent*. For example, the article in Fig. 1 would be labeled as related because both the headline and the body text cover the topic of yoga; however, the headline and the body text are incongruent.

This current study tackles the headline incongruence problem [7], which is a significant kind of misinformation that originates from a discrepancy in a news headline and the corresponding body text. No million-scale dataset has been openly available for this problem.

2.2 Designing Web Interfaces for News Readers

A line of research was conducted in which a news service was constructed as a separate system. A decade ago, a pioneering study provided a news service on the web [30]. The researchers designed NewsCube, which was a news service that aimed at mitigating the effect of media bias. The service provided readers with multiple classified viewpoints on a news event of interest, which facilitated the formulation of more balanced views. Most recently, a study implemented a web system that highlights objective sentences in a user text to mitigate the biased reporting that facilitates polarization [26]. Another study developed a visualization tool that enables Twitter users to explore the politically-active parts of their social networks [18] and conducted a randomized trial to evaluate the impact of recommending accounts of the opposite political ideology. The construction of a separate news service enables researchers to investigate the effects of a machine while controlling for other factors. However, it mostly serves as a proof of concept; hence, the impact on end-users is limited.

Many stand-alone systems suffer from gaining traffic. Therefore, other studies have developed a lightweight browser widget that operates with available news services on the web, which enables more users to be reached in practice. One study presented a browser widget that encourages the reading of diverse political viewpoints based on the selective exposure theory [28]. According to a field deployment study, showing feedback led to more balanced exposure. A browser extension was also implemented to help people determine whether a headline is clickbait or a general headline [4]; however, the effects of this mechanism were not evaluated.

Motivated by these studies that employ browser widgets, this study implements a lightweight web interface that helps readers determine whether a specified headline is incongruent before clicking on the headline. We also conduct a user study via questionnaires and in-depth interviews to estimate the impact of the web interface.

3   Data and Methodology

This section presents the approach to building a million-scale dataset for the headline incongruence problem and the methodology for detecting such misleading headlines via neural networks. The objective is to determine whether a news article contains an incongruent headline, given a pair that consists of a headline and a body text. For detection models, we call the output probability of being an incongruent headline the **incongruence score**.

3.1   Dataset Generation

One natural method for constructing a labeled dataset is for researchers and crowd-sourced workers to manually annotate data. However, the training of sophisticated classification algorithms requires a large dataset, which is not feasible to obtain via manual annotation due to high cost and reliability issues. Alternatively, this work presents a systematic approach for the automatic generation of million-scale datasets that are composed of incongruent and correct headlines.

First, we crawled a nearly complete set of news articles that were published in South Korea from January 2016 to October of 2017. From over 4 million news articles, we conducted a series of cleansing steps, such as removal of noncritical information (e.g., reporter name and nontextual information such as photos and videos). Next, we transformed word tokens to integers, which is released with vocabulary to help researchers utilize the dataset without being hindered by a language barrier.

To label the incongruity of headlines for millions of news articles, we implanted unrelated or topically inconsistent content into the body text of original news articles rather than crafting new headlines. Figure 2 illustrates the generation process of

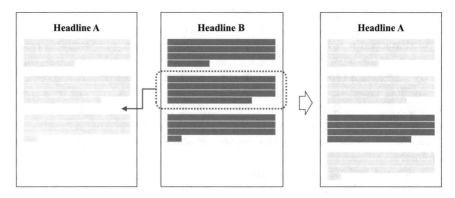

Fig. 2 An illustration on the generating process of incongruent headlines

incongruent headlines. This process can produce a pair that consists of a headline and a body text such that the headline tells stories that differ from the body text content. Hence, the automation process for creating incongruent-labeled data involves the following steps: (1) sampling a target article from the corpora, (2) sampling part-of-content from another article of the corpora, and (3) inserting this part-of-content into the target article. We controlled the topics of the sampled paragraphs to be similar to each target article by employing the meta-information on news articles (e.g., news category).

We created the congruent-labeled data by selecting them from suitable corpora. No headline in this set overlaps with the incongruent-labeled data. Nonetheless, this process may yield false-negative instances if a real article that has an incongruent headline is chosen inappropriately as a target. We conducted additional steps to reduce the number of false negatives via rule-based processing, such as the inspection of advertising phrases with an n-gram dictionary. We also hired human annotators to read 1,000 randomly sampled articles from the created dataset and to check whether those articles are labeled correctly. These efforts minimize the number of errors that can arise from the automatic generation process. The final corpus is composed of a training set of 1.7 M news articles that are balanced against the incongruity label. For evaluation, we maintained separate development and test sets of 0.1 M instances each. The statistics of the datasets are presented in Table 1.

This approach is language-agnostic; hence, it can be applied to any news corpora of any language. The generated dataset is publicly available on the GitHub page.[1]

3.2 Baseline Approaches

We introduce four baseline approaches that have been applied to the headline incongruence problem. Feature-based ensemble algorithms have been widely utilized for their simplicity and effectiveness. Among various methods, the XGBoost algorithm has shown superior performance across various prediction tasks [5]. For example, in a recent challenge on determining the stances of news articles [13], the winning team applied this algorithm based on multiple features to measure similarities between the headline and body text [40]. As a baseline, we implemented the **XGBoost (XGB)** classifier by utilizing the set of features that are described in the winning model,

Table 1 Dataset statistics

	Mean	Std. Error
Number of tokens in headline	13.71	0.003
Number of tokens in body text	513.97	0.208
Number of paragraphs in body text	8.17	0.004
Number of tokens in paragraph	61.7	0.018

---

[1] http://github.com/david-yoon/detecting-incongruity/

such as cosine similarities between the headline and body text. In addition to this model, we trained **support vector machine (SVM)** classifiers based on the same set of features.

Recurrent Dual Encoder (RDE) A recurrent dual encoder that consists of dual Recurrent Neural Networks (RNN) has been utilized to calculate a similarity between two text inputs [27]. We apply this model to the headline incongruence problem via dual RNNs based on gated recurrent unit (GRU) that encode the headline and body, respectively. When a RNN encodes word sequences, each word is passed through a word-embedding layer that converts a word index to a corresponding 300-dimensional vector. After the encoding step, the probability of being incongruent headline is calculated by using the final hidden states of RNNs for headline and body text. The incongruence score in the training objective is as follows:

$$p(\text{label}) = \sigma((h_{t_h}^H)^\mathsf{T} M \, h_{t_b}^B + b),$$

$$\mathcal{L} = -\log \prod_{n=1}^{N} p(\text{label}_n | h_{n,t_h}^H, h_{n,t_b}^B), \tag{1}$$

where $h_{t_h}^H$ and $h_{t_b}^B$ are last hidden state of each headline and body text RNN with the dimensionality $h \in \mathbb{R}^d$. The $M \in \mathbb{R}^{d \times d}$ and bias b are learned model parameters. N is the total number of samples used in training and σ is the sigmoid function.

Convolution Dual Encoder (CDE) Following the Convolutional Neural Network (CNN) architecture for text understanding [23], we apply Convolutional Dual Encoder to the headline incongruence problem. Taking the word sequence of headline and body text as input to the convolutional layer, we obtained a vector representation $v = \{v_i | i = 1, \cdots, k\}$ for each part of the article through the max-over-time pooling after computing convolution with k filters as follows:

$$v_i = g(f_i(W)), \tag{2}$$

where g is max-over-time pooling function, f_i is the CNN function with i-th convolutional filter, and $W \in \mathbb{R}^{t \times d}$ is a matrix of the word sequence. We use dual CNNs to encode a pair of headline and body text into vector representations. After encoding each part of a news article, the probability that a given article has the incongruent headline is calculated in a similar way to the Eq. (1).

3.3 Proposed Methods

While the available approaches perform reasonably for short text data, dealing with a long sequence of words in news articles will result in degraded performance [31].

For example, RNN that is utilized in RDE performs poorly in remembering information from the distant past. While CDE learns local dependencies between words, the typical short length of its convolutional filter prevents the model from capturing any relationships between words in distinct positions. The inability to handle long sequences is a critical drawback of the standard deep approaches to the headline incongruence problem because a news article can be very long. As presented in Table 1, the average word count per article is 513.97 in our dataset.

Therefore, we fill this gap by proposing neural architectures that efficiently learn hierarchical structures of long text sequences. We also present a data augmentation method that efficiently reduces the length of the target content while increasing the size of the training set.

Hierarchical Recurrent Dual Encoder (HRDE) Inspired by previous approaches that models text using a hierachical architecture [44–46], this model divides the text into a list of paragraphs and encodes the entire text input from the word level to the paragraph level using a two-level hierarchy of RNN architectures.

For each paragraph, the word-level RNN encodes the word sequences $w_p = \{w_{p,1:t}\}$ to $h_p = \{h_{p,1:t}\}$. Next, the hidden states of the word-level RNN are fed into the next-level RNN that models a sequence of paragraphs while preserving the order. The hierarchical architecture can learn textual patterns of news articles with fewer sequential steps for RNNs compared to the steps required for RDE. While RDE involves an average of 513.97 steps to learn news articles in our dataset, AHDE only accounts for 61.7 and 8.17 steps on average for word- and paragraph-level of RNN, respectively (see Table 1). The hidden states of hierarchical RNNs are as follows:

$$
\begin{aligned}
h_{p,t} &= f_\theta(h_{p,t-1}, w_{p,t}), \\
u_p &= g_\theta(u_{p-1}, h_p),
\end{aligned}
\tag{3}
$$

where u_p is the hidden state of the paragraph-level RNN at the p-th paragraph sequence, and h_p is the word-level RNN's last hidden state of each paragraph $h_p \in \{h_{1:p,t}\}$. We use the same training objective as the RDE model such that the incongruence score, the probability of having incongruent headlines, is calculated as follows:

$$
p(\text{label}) = \sigma((u_{p_h}^H)^\mathsf{T} M\, u_{p_b}^B + b)
\tag{4}
$$

Attentive Hierarchical Dual Encoder (AHDE) In addition to the hierarchical architecture of HRDE, attention mechanism is employed to the paragraph-level RNN to enable the model to learn the importance of each paragraph in a body text for detecting incongruity embedded in the corresponding headline. Additionally, we utilize bi-directional RNNs for the paragraph-level RNN to learn sequential information in both directions from the first paragraph and the last paragraph.

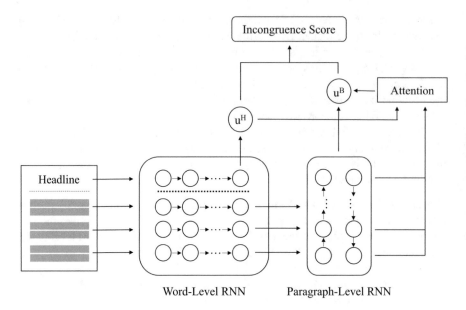

Fig. 3 Diagram of AHDE. Every paragraph is encoded into a hidden state, and the sequence of the hidden states corresponding to each paragraph is further encoded into the hidden state corresponding to the entire body text. The model can learn importance of paragraphs in a body text for detecting headline incongruity from an attention mechanism

Figure 3 illustrates a diagram of AHDE. Each u_p of a body text is aggregated according to its correspondence with the headline as follows:

$$s_p = v^\mathsf{T} tanh(W_u^B u_p^B + W_u^H u^H),$$
$$a_i = \exp(s_i)/\textstyle\sum_p \exp(s_p), \tag{5}$$
$$u^B = \textstyle\sum_i a_i u_i^B,$$

where u_p^B indicates the p-th hidden state of the paragraph-level RNN that learns the representation of a body text. The u^H indicates the last hidden state of the paragraph-level RNN with the corresponding headline. Similar to HRDE, the incongruence score is calculated as follows:

$$p(\text{label}) = \sigma((u^H)^\mathsf{T} M \, u^B + b) \tag{6}$$

Hierarchical Recurrent Encoder (HRE) The HRDE and AHDE model uses two hierarchical RNNs for encoding text from the word level to the paragraph level. Compared to non-hierarchical alternatives such as RDE and CDE, those models require higher computation resources in training and inference. Therefore, we investigate a moderate approach that models hierarchical structures of news articles with a simpler neural architecture. A body text is divided into paragraphs, each of which is represented by averaging word-embedding vectors of words within the paragraph. In other words, HRE calculates h_p in Eq. (3) by the average of the word vectors in the paragraph p, $h_p = \sum_i embedding(w_i)$, $w_i \subset p$-th paragraph. Then, a paragraph-level RNN is applied to the paragraph-encoded sequence input, h_p, for retrieving the final encoding vector of the entire body text. The incongruence score is calculated by

$$p(\text{label}) = \sigma((h^H)^\mathsf{T} M \, h_p + b) \qquad (7)$$

where h^H indicates the average embedding vector of the words in the headline.

Independent Paragraph (IP) method In addition to the neural architecture, we propose a data augmentation method that divides a body text into multiple paragraphs and learns the relationship between each paragraph and the corresponding headline independently. For that purpose, we transformed every pair that consists of a headline and a body text in the original dataset into multiple pairs of the headline and each paragraph. This conversion process not only reduces the length of text that a model should process but also increase the total number of training instances. For example, the average number of words in a body text shrinks from 513.97 to 61.7 (see Table 1), and the number of training instances increases from 1.7 M to 14.2 M. We expect that this difference makes the proposed deep learning models efficiently learn the pattern of the semantic mismatch between a headline and its body text.

Figure 4 illustrates the diagram of the IP method, which computes **incongruence score** of each paragraph from its relationship with the corresponding headline. The final incongruence score for the pair of the headline and the body text is determined as the maximum of the incongruence score for the headline and each paragraph as follows:

$$p(\text{label}) = max(s_{1:p}), \qquad (8)$$

where s_p is the incongruence score calculated from the p-th paragraph of the body text and the headline. The selection of the maximum score can better identify news articles that contain a paragraph that is highly unrelated to the news headline. We

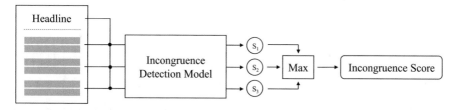

Fig. 4 Diagram of the IP method. A body text is divided into multiple paragraphs, each of which is compared to the corresponding headline to calculate the incongruence score of each paragraph. The maximum value of all scores is the incongruence score for the pair of the headline and the body text

also tested other aggregation methods such as average and minimum, but max function led to the best performance.

With the use of the IP method, hierarchical approaches consider sentence in a paragraph the lower unit in two-level hierarchy of neural architectures. In particular, the incongruence score of each detection model is calculated in the following ways:

- **XGB/SVM with IP:** For each paragraph, XGB/SVM measures the incongruence score by extracting features from its headline and the paragraph.
- **RDE/CDE with IP:** Both models encode word sequences in each paragraph of a body text and compare them with the corresponding headline.
- **HRDE/AHDE with IP:** To obtain the incongruence score for each paragraph, the first-level RNN encodes word sequences for each sentence in the paragraph, and the second-level RNN takes as input the hidden states of the sentences that are retrieved from the first-level RNN.
- **HRE with IP:** HRE calculates the mean of word vectors for each sentence. Then, a RNN encodes a sequence of sentences by taking the averaged word vectors as input.

4 Evaluation Experiments

4.1 Automatic Evaluation

Table 2 presents the performances of all approaches on the test set. We report the accuracy and the AUROC (area under the receiver operating characteristic curve) value, which is a balanced metric for the label distribution. Here, we make three main observations.

First, among the baseline models, RDE realized the best performance with an accuracy of 0.845 and an AUROC of 0.939. The decent performance of RDE suggests that recurrent neural networks are well suited to the learning of sequential text representations of news articles, in contrast to the feature-based approaches and the convolutional encoders, which learn the local dependencies of word tokens.

Table 2 Model performance with and without the Independent Paragraph (IP) method. Top-2 scores are marked as bold. The top 4 rows indicate the baseline performance and the bottom 3 rows shows the performance of the proposed models

Model	Without IP		With IP	
	Accuracy	AUROC	Accuracy	AUROC
SVM	0.640	0.703	0.677	0.809
XGB	0.677	0.766	0.729	0.846
CDE	0.812	0.9	0.870	0.959
RDE	0.845	0.939	0.863	0.955
HRDE	0.885	0.944	0.881	0.962
AHDE	0.904	0.959	0.895	0.977
HRE	0.85	0.927	0.873	0.952

Second, the performance margin increased significantly when hierarchical structures were applied to RDE. In HRDE, the accuracy and AUROC increased by 0.04 and 0.05, respectively. Considering the hierarchical structure of news articles in the design of neural architectures may facilitate the learning of textual information of news articles more efficiently, such that headline incongruity can be more accurately identified. In contrast, in HRE, merely inputting the mean word representation for each paragraph into a single layer recurrent network did not yield a significant improvement. Compared to RDE, the accuracy increased with a margin of 0.005; however, the AUROC decreased by 0.012. Third, we found the attention ability of AHDE further enhanced the performance up to an accuracy of 0.904 and an AUROC of 0.959, namely, knowledge of relevant paragraphs in the detection of incongruent headlines facilitated the efficient examination of the relationship between the headline and each paragraph by AHDE.

Last, the prediction performance increased significantly when the IP method was applied. RDE and CDE benefitted most from the application of the IP method; they even showed performances that were comparable to those of the hierarchical models. Although those simple models do not have a suitable structure for handling lengthy news data (on average, the body texts and the paragraphs contain 518.97 and 61.7 words, respectively, according to Table 1), the IP method helped them examine the relationship between the headline and each paragraph more efficiently.

4.2   Manual Evaluation

To test the efficacies of the dataset and the proposed models for the detection of incongruent headlines in the wild, we evaluated the pretrained models on more recently published news articles. We gathered 232,261 news articles that were published from January to April of 2018. Via evaluation of the model on this recently assembled dataset, we can measure the generalizability of our approaches to dataset generation and headline incongruity detection in practice.

First, we manually inspected random samples of news articles to determine whether they have incongruent headlines; however, we could not retrieve sufficiently many instances with incongruent headlines for evaluation. The lack of misleading articles is possibly due to the sparsity of such headlines in practice, despite their critical importance. Therefore, instead of manually labeling randomly sampled news articles, a majority of which may correspond to general headlines, we manually evaluated the top-N articles in terms of the incongruence scores that are assigned by our model. Since models assign incongruence scores (output probabilities) based on their confidence for classification, we believe such evaluation successfully estimates the degree of precision of a prediction model. This type of assessment is widely used in tasks in which it is impossible to count all possible real cases in a dataset such as question answering system [15].

Figure 5 presents the precision scores for the AHDE models that are trained with and without the IP method. The x-axis corresponds to the top-N articles in terms of

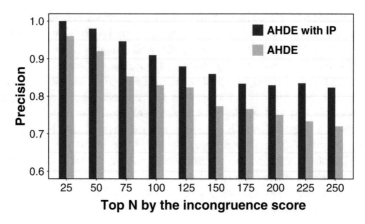

Fig. 5 Precision values for detecting news articles with incongruent headlines in the newly gathered dataset. The x-axis shows the top-N articles by incongruence scores, and the y-axis presents its corresponding precision

the incongruence scores that are assigned by the models out of the newly gathered news articles over 4 months. The y-axis corresponds to the precision values of the top-N articles. Here, we make three observations.

First, the AHDE model with the IP augmentation consistently shows higher precision than the AHDE model without the IP method. This finding supports the superior performance of the IP method across evaluations. Second, the AHDE model with the IP method realized a precision of 1.0 for the top 25 articles. Even though the model was trained on a separate dataset, it successfully filtered out real cases in which the headline conveys different information than the associated body text. Third, when we evaluate the top 250 articles, the precision of the AHDE model with the IP method reduced to 0.82. This precision value is sufficiently high for the detection of incongruent headlines in real news platforms.

5   BaitWatcher: A Lightweight Web Interface for the Detection of Headline Incongruity

This section introduces a new web interface that aims at reducing the adverse effects of incongruent news headlines on the news reading experience. Incongruent headlines can mislead readers with an unexpected body text because they are one of the critical cues that are used in the selection of news articles in online environments. Before clicking on a headline and reading the body text, newsreaders are not able to determine in advance the content of the news story. We hypothesize that news readers will be empowered if they are given a choice and additional information about the headline incongruence score. As a proof of concept, we designed and implemented a web interface, namely, *BaitWatcher*, that quickly

Fig. 6 The user interface of BaitWatcher

reports the incongruence score. We conducted a focus group interview to investigate the effects of the web interface.

5.1   Design and Implementation

The main feature of BaitWatcher is that it reveals the likelihood of a specified news headline being incongruent to its full body text based on the presented deep learning model. This information is made visible to users *before* they click on the headline to read the entire story. BaitWatcher is platform-agnostic and can be implemented on top of any news platforms. We expect that revealing the hidden information through a simple interface will empower news readers by helping them determine by themselves whether to read news articles with potentially incorrect headlines or not. As shown in Fig. 6, if a user hovers a mouse pointer over a news headline of interest, the BaitWatcher interface immediately displays the prediction result (the sigmoid output) of a pretrained deep neural network via a tooltip view. This additional information facilitates readers in the selection of which news articles to read in detail. Once a user decides to read a news article, the full news story will be made available to them as usual, along with a user feedback section that appears at the bottom of the page. This feedback section was implemented in the form of a button that signals whether the news story was consistent with the context that was provided by the news headline. This process enables the system to gather manual labels on incongruent news articles in the wild, which will be used to train the deep learning model periodically to increase the accuracy and robustness of detection.

To reduce the computational burden of running a deep model, BaitWatcher was implemented as a browser extension that is based on a client-server architecture. After installing the extension on a browser (e.g., Chrome), online users can choose to read news articles on any news platforms and obtain information about the incongruence scores of news headlines before reading the full corresponding body text. As shown in the left side of Fig. 6, if a reader hovers the mouse cursor over a news headline, BaitWatcher sends an HTTP request from the client to the

API server with the hyperlink on the article. The server parses the news content via the Python Newspaper3k library,[2] which uses advanced algorithms with web scrapping to extract and parse online newspaper articles. The parsed content is fed into the pretrained deep learning model to return the incongruence score. The AHDE model with IP was selected as the model since this algorithm realized the best performance in the evaluation experiment. Because the Python Newspaper3k library automatically detects headline and body text, BaitWatcher can be run on any news website. The code and implementation details are publicly available on the GitHub page.[3]

5.2 Focus Group Interview

After implementing BaitWatcher, we evaluated its performance in a realistic setting. We conducted a small-scale focus group interview to gain insight into how the provision of additional information about headline incongruence can improve the news reading experience. A total of fourteen participants of ages 20–29 were recruited from the second author's institute, all of whom identified themselves as moderate to avid news consumers. All participants said they actively read news articles at least once a week. After hearing a brief introduction to BaitWatcher's functionality, each focus group participant was given 30 min to read news articles through the BaitWatcher interface. While BaitWatcher can be deployed on any news website as discussed earlier, we asked the focus group participants to visit a common news portal for finding news [29] to minimize the effects of distinct media outlets on the perception of headline incongruence. After the 30 min news reading experience in the lab, we conducted an open interview with each participant. Institutional Review Board had approved this focus group survey and the news assistant experimental design at the second author's institute (Approval code: #KH2018-62).

The open interview included the following questions, which capture the news reading habits of users and quantify the effectiveness of BaitWatcher:

Q1. How often do you read news online in a week?
Q2. Which category of news are you mostly interested in?
Q3. When you are reading news online, how likely are you to read the full story?
Q4. Does showing incongruence scores affect the choice of news articles to read?

Table 3 displays the necessary information about the participants and the questionnaire results. Here, we make observations on their reading behaviors and the effects of BaitWatcher in preventing readers from clicking on incongruent headlines. First, as previous studies noted [12, 16], a significant degree of participants (78.5%)

---

[2]https://newspaper.readthedocs.io/

[3]https://github.com/bywords/BaitWatcher

Table 3 Participants' information and questionnaire results

Participant	Gender	Age	Q1 (Freq)	Q2 (Interests)	Q3 (Full story)	Q4 (BaitWatcher)
P1	Male	24	7 days	Politics	Less likely	No
P2	Male	23	7 days	Politics	Less likely	No
P3	Male	21	3–4 days	Politics	Less likely	Yes
P4	Female	22	2 days	Entertainment	Less likely	Yes
P5	Female	20	7 days	Politics	More likely	Yes
P6	Female	22	7 days	Social issues	More likely	Yes
P7	Male	22	5 days	Social issues	Less likely	Yes
P8	Male	20	4 days	Sports	Less likely	Yes
P9	Male	26	3–4 days	Life & Culture	Less likely	Yes
P10	Male	26	7 days	Economics	More likely	Yes
P11	Female	24	7 days	Entertainment	Less likely	No
P12	Female	24	2 days	IT & Science	More likely	No
P13	Female	24	7 days	Sports	Less likely	Yes
P14	Male	28	7 days	Politics	Less likely	Yes

reported that they are more likely to consume headlines without reading the full news stories. This skimming behavior may enable them to browse a more extensive set of news stories every day; however, it makes them vulnerable to misleading headlines such as clickbait and incongruent headlines. This result supports the necessity of showing the incongruence score before the user clicks on a headline.

In response to the question on the effects of BaitWatcher (Q4), ten out of fourteen participants (71.4%) reported that the use of this interface affected their choices of news articles to read, whereas four participants (28.6%) responded that they were not influenced by or did not benefit from this web interface. Particularly, *P12*, whose interest is in reading 'IT & Science' news, responded 'No' to this question because the participant did not encounter any news stories for which the incongruence score was alarmingly high within the set of news stories that were browsed. Therefore, the user could not experience the benefits of BaitWatcher. The frequency of incongruent headlines is typically low in practice and can vary across topics. Nonetheless, according to a more significant proportion of the participants, having this additional information seems useful and empowering.

Those who answered 'Yes' to Q4 reported that BaitWatcher was "interesting" and "effective" in that they avoided clicking on news headlines with high incongruence scores, as we had hypothesized. Three participants (P4, P8, and P14) mentioned that they were attracted to such incongruent headlines because they wanted to inspect the articles that BaitWatcher reported to be incongruent to their headlines.

> (P4) "... At first, I became curious about why certain headlines were labeled as incongruent by BaitWatcher, so I clicked on them and checked how the articles looked ... "

The unexpected browsing behaviors support the findings of previous studies on the adverse effects of labeling on the prevention of fake news [9, 17]. From the opinions of two participants (P8 and P9), we identify new potential to mitigate the unnecessary attention that high incongruence scores receive. A possible strategy is to pursue the interpretability of the deep learning models and to present the results as grounds for the high score and how the algorithm works. When an algorithm looks like a black box, users will naturally question its prediction results. Another strategy is to present ample examples of news articles with high incongruence scores in advance of the experiment to facilitate understanding of the participants regarding the general performance of BaitWatcher.

> (P8) "... When BaitWatcher displayed a high score, it made me wonder, "why does this headline have such a high score?" This led me to click the headline and guess the reasoning that the AI used in making the decision..."

> (P9) "... I did not click incongruent headlines because BaitWatcher warned me not to do so. Nonetheless, whenever it (BaitWatcher) showed high scores, I was curious why the AI made such a decision. It may be effective for people to see the internal reasoning process of this AI model ..."

Overall, our focus group study demonstrated that the provision of the incongruence score in today's news reading is empowering to users. Web interfaces such as BaitWatcher will not only prevent newsreaders from clicking on news headlines that are likely incongruent to their full linked stories but also gradually build people's trust over time in the model's predictions. Whether one is an avid news reader or not, spending time on incongruent stories is an unpleasant experience in most cases. A headline might be deliberately misleading due to sarcasm, in which case readers could still click on the news article and enjoy reading it even if BaitWatcher's reported score is high. The deep models that are proposed in this work do not yet provide high interpretability, and detection models that are also interpretable could be developed in future studies.

6 Discussion

The role of headlines in the news reading experience has been studied in journalism and communication research. News headlines should provide a concise and accurate summary of the main story, thereby enabling readers to decide whether to read the news story [38]. Online social media and the web have become convenient platforms for news consumption. According to Digital News Reports by Reuters Institute [33], a third fourth of the survey participants replied that they consume news through online media. In contrast to news consumption via traditional outlets such as newspapers, the main content is not shown to readers in online media; only headlines and visual snippets are exposed. Hence, newsreaders are more likely to consume only the news headlines and not the full news stories—a behavior that some refer to as *a shopper of headlines* [12]. In such environments, if a news headline does not accurately represent the story, it could mislead readers into disseminating false information [2, 41], which could lead to pressing social problems. Even though the proportion of incongruent headlines is not large against the numerous news articles that are published each day, an inaccurate impression can percolate through a user's online networks and eventually lead to severe social problems such as polarization, as a previous study similarly discovered in the context of fake news consumption on Twitter [20]. This study identified the dangers and problems that are associated with headline-led news reading, and its contributions are three-fold.

First, we release a dataset of 1.7 million news articles that are constructed on the entire articles published in a nation over two years. Due to the sparsity of incongruent headlines in the wild, it requires a considerable amount of time and effort to develop a sizable and balanced dataset via manual annotation. Therefore, previous studies introduced small datasets that are not suitable for training sophisticated models. To address the issue of scalability in the construction of a dataset, we automatically generate incongruent headlines by implanting paragraphs of other articles into the body text. This generation process can be applied to any set of news articles in any language, which will facilitate future studies on the application of data-driven approaches to incongruent headlines.

Second, this study proposes an attention-based hierarchical neural network for the headline incongruence problem. While recurrent neural networks are efficient in modeling sequential information such as text, a body text hinders the propagation of error signals via backpropagation. Thus, inspired by the hierarchical structure of a news article that is composed of paragraphs, we design a hierarchical recurrent network that models word sequences of each paragraph into a hidden state and combines the sequence of the paragraphs through another level of the recurrent neural network. This newly proposed model outperformed baseline approaches with an AUROC of 0.977 on the detection task.

Third, we implement a lightweight web interface that facilitates the selection by readers of relevant articles to read in a typical scenario of online news consumption in which only headlines are shown. The results of a focus group interview demonstrate the effectiveness of the interface in preventing users from selecting those articles and suggest a future direction for the improvement of deep learning models. Similar to the findings of a recent work [21], the participants require a high level of interpretability on model predictions, which is not embedded in the proposed models. Following the recent efforts on deep learning [36], the development of an interpretable model will help build a high level of trust in machine-based decisions on incongruent headlines, which will be crucial for the utilization of such interfaces in practice.

6.1 Hierarchical Encoders for Stance Detection

To further evaluate the generalizability of the deep approaches that are proposed in this paper, we conducted an additional experiment on the FNC-1 dataset [13], with the objective of *stance detection*. This problem is similar to the headline incongruence problem in that one must compare the textual relationships between news headlines and the corresponding full content but different in that its target label consists of four separate cases (unrelated, agree, disagree, and discuss). To obtain a similar setting to that of our task, we transformed these four labels into binary labels: "unrelated" and "others".

We compared our hierarchical deep learning approaches (AHDE, HRDE, and HRE) with feature-based methods and standard deep learning models. We also considered ensemble models that combine the predictions of XGB and each deep learning model, because an ensemble of XGB and CDE was the winning model of the FNC-1 challenge [40]. XGB outperformed the other single models with an accuracy of 0.9279. Among deep learning models, the AHDE model realized the highest accuracy of 0.8444. The superior performance of XGB over deep approaches might result from insufficient variations among the training instances in the FNC-1 dataset. Even though the training set contains approximately 50 K examples, many news articles that correspond to the independent label were generated from 1,683 original news articles by swapping headlines with one another; thus, 29.7 cases had identical body text.

These reasons might have led the challenge winners to use ensemble models that combine the predictions of feature-based approaches and deep neural networks. The XGB+CDE ensemble model realized the accuracy of 0.9304 and outperformed all the single models. When we combined the predictions of AHDE with XGB, the ensemble model produced the best accuracy of 0.9433. Incorporating the results in Table 2, this finding suggests that the proposed hierarchical neural networks effectively learn textual relationships between two texts in contrast to standard approaches. We firmly believe that the highest accuracy of XGB among the single models is due to the limitation of the FNC-1 dataset, as discussed earlier; hence, the ensemble approach may not be necessary if the dataset is sufficiently large for neural network training. According to additional experiments on the dataset that was proposed in this paper, the AHDE model outperformed all combinations of other approaches for the ensemble.

6.2 Varying Perceptions on Headline Incongruence

So far, we have treated the incongruent score as an inherent value that is fixed for each news article. We conducted additional surveys using the Amazon Mechanical Turk (MTurk) service to determine whether the general public would agree with the predictions by our models regarding which news articles contain incongruent headlines. We also evaluated whether people's perceptions of incongruence scores vary according to their partisanship, and we hypothesize based on a previous finding that people's knowledge of the veracity of news varies by political stance [1].

First, we manually gathered news articles from two media outlets. To retrieve as many incongruent news headlines as possible, we selected two media outlets that are considered not trustworthy by common journalistic standards (referring to mediabiasfactcheck.com): one was chosen from the conservative media (**Media A**) and another from the liberal media (we call **Media B**). We do not reveal these media names, as the choices of media outlets are less of a concern in our study.

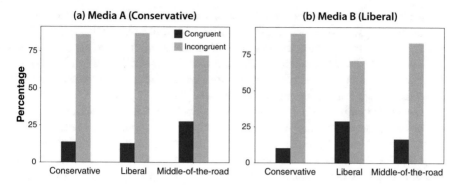

Fig. 7 MTurk results indicating political stances of survey participants (the x-axis) and their responses to articles of high incongruence score (the y-axis). (**a**) indicates a media platform with a Conservative bias; (**b**) indicates a media platform with a Liberal bias

Given the definition of an incongruent headline and the incongruent articles that are selected by the model from each media outlet, we asked 100 Amazon Mechanical Turk workers to answer the following question *"Do you think the headline of the above article is incongruent with its body text?"*

According to Fig. 7, MTurk workers tend to find that articles with high incongruence scores contain misleading headlines. One exciting trend is the dependence of the perceived incongruence score on individual belief. While nonliberal participants considered news samples from **Media B** to have a similar level of incongruence to samples from **Media A**, liberal participants found **Media B** to be less incongruent. This finding suggests that while our approach is applicable in general scenarios, the perceived incongruence level may be judged differently among news topics (such as politics). News service providers and researchers should be cautious when employing human coders and crowdsourcing workforces to obtain fair labels on misinformation and fake news.

6.3   Future Directions

A natural extension of this study is the development and improvement of prediction models for detecting news articles with incongruent headlines by incorporating NLP techniques with deep learning approaches. For example, one could apply named entity recognition as a preprocessing step to represent word tokens in an embedding space better. It would also be possible to consider syntactic features in modeling text by developing tree-shaped deep neural architectures that are similar to LSTM-tree [39].

Another future direction is to devise a learning-based approach for generating headlines. To construct a million-scale dataset for training incongruity detection models, we modified the body text while keeping the original news headline unchanged. While the process has shown to generate a training corpus effectively, researchers could develop an AI agent that rewrites a headline that is incongruent with its body text. While the research on text generation has lagged behind the remarkable achievements in image domains due to the difficulty of handling discrete outputs, future research could be extended from recent studies on controlled generation [22] or cross-alignment style transfer [37].

Beyond the online news domain, this work could lead to new measurements of the incongruence of title and content across other types of online content. The title plays a crucial role in enticing users to click and consume digital content such as blog articles, online videos, and even scientific papers. Similar to the incongruent headline problem, the automatic identification of such incongruent titles of various content will improve people's online experiences. Future researchers could share multiple types of datasets and could develop AI approaches that measure the inconsistency between title and content.

7   Conclusions

Here, we study the detection of incongruent headlines that make claims that are unsupported by the corresponding body texts. We release a million-scale data corpus that is suitable for the detection of the misleading headline. We also propose deep neural networks that efficiently learn the textual relationship between headline and body text via a recurrent hierarchical architecture. To further facilitate news reading in practice, we present BaitWatcher, which is a lightweight web interface that presents to readers the prediction results that are based on deep learning models before the readers click on news headlines. The code and implementation details are released for broader use, and we hope this study contributes to the construction of a more trustworthy online environment for reading news.

Acknowledgements This research was supported by Basic Science Research Program through the National Research Foundation of Korea (NRF) funded by the Ministry of Science and ICT (No. NRF-2017R1E1A1A01076400).

References

1. Allcott, H., Gentzkow, M.: Social media and fake news in the 2016 election. J. Econ. Perspect. **31**(2), 211–36 (2017)
2. Allport, F.H., Lepkin, M.: Building war morale with news-headlines. Public Opin. Q. **7**(2), 211–221 (1943)
3. Blom, J.N., Hansen, K.R.: Click bait: forward-reference as lure in online news headlines. J. Pragmat. **76**, 87–100 (2015)
4. Chakraborty, A., Paranjape, B., Kakarla, S., Ganguly, N.: Stop clickbait: detecting and preventing clickbaits in online news media. In: Proceedings of the ASONAM (2016)

5. Chen, T., Guestrin, C.: Xgboost: a scalable tree boosting system. In: Proceedings of the KDD (2016)

6. Chen, Y., Conroy, N.J., Rubin, V.L.: Misleading online content: recognizing clickbait as false news. In: Proceedings of the ACM Workshop on Multimodal Deception Detection (2015)

7. Chesney, S., Liakata, M., Poesio, M., Purver, M.: Incongruent Headlines: yet another way to mislead your readers. In: Proceedings of the EMNLP Workshop, pp. 56–61 (2017)

8. Chopra, S., Jain, S., Sholar, J.M.: Towards automatic identification of fake news: Headline-article stance detection with LSTM attention models. In Stanford CS224d Deep Learning for NLP final project (2017)

9. Clayton, K., Blair, S., Busam, J.A., Forstner, S., Glance, J., Green, G., Kawata, A., Kovvuri, A., Martin, J., Morgan, E.: Real Solutions for fake news? Measuring the effectiveness of general warnings and fact-check tags in reducing belief in false stories on social media. Polit. Behav. 1–23 (2019)

10. Clickbait Challenge. http://www.clickbait-challenge.org (2017). [Online; Accessed 15 April 2019]

11. Ecker, U.K., Lewandowsky, S., Chang, E.P., Pillai, R.: The effects of subtle misinformation in news headlines. J. Exp. Psychol. Appl. 20(4), 323 (2014)

12. English, E.: A study of the readability of four newspaper headline types. Journal. Bull. 21(3), 217–229 (1944)

13. Fake News Challenge. http://www.fakenewschallenge.org/ (2017). [Online; Accessed 15 April 2019]

14. Ferreira, W., Vlachos, A.: Emergent: a novel data-set for stance classification. In: Proceedings of the NAACL-HLT (2016)

15. Ferrucci, D.A.: Introduction to "this is Watson". IBM J. Res. Dev. 56(3–4), 1–1 (2012)

16. Gabielkov, M., Ramachandran, A., Chaintreau, A., Legout, A.: Social clicks: what and who gets read on Twitter? ACM SIGMETRICS Perform. Eval. Rev. 44(1), 179–192 (2016)

17. Gao, M., Xiao, Z., Karahalios, K., Fu, W.T.: To label or not to label: the effect of stance and credibility labels on readers' selection and perception of news articles. Proc. ACM Hum. Comput. Interact. 2(CSCW), 55 (2018)

18. Gillani, N., Yuan, A., Saveski, M., Vosoughi, S., Roy, D.: Me, my echo chamber, and i: introspection on social media polarization. In: Proceedings of the WWW, pp. 823–831 (2018)

19. Go, A., Bhayani, R., Huang, L.: Twitter sentiment classification using distant supervision. CS224N project report, Stanford, 1(12), (2009)

20. Grinberg, N., Joseph, K., Friedland, L., Swire-Thompson, B., Lazer, D.: Fake news on Twitter during the 2016 US presidential election. Science 363(6425), 374–378 (2019)

21. Horne, B.D., Nevo, D., O'Donovan, J., Cho, J.H., Adali, S.: Rating reliability and bias in news articles: does AI assistance help everyone? arXiv preprint arXiv:1904.01531 (2019)

22. Hu, Z., Yang, Z., Liang, X., Salakhutdinov, R., Xing, E.P.: Toward controlled generation of text. In: Proceedings of the ICML, pp. 1587–1596 (2017)

23. Kim, Y.: Convolutional neural networks for sentence classification. In: Proceedings of the EMNLP, pp. 1746–1751 (2014)

24. Kuiken, J., Schuth, A., Spitters, M., Marx, M.: Effective headlines of newspaper articles in a digital environment. Digit. Journal. 5(10), 1300–1314 (2017)

25. Kwon, S., Cha, M., Jung, K., Chen, W., Wang, Y.: Prominent features of rumor propagation in online social media. In: Proceedings of the ICDM (2013)

26. Lovering, C., Lu, A., Nguyen, C., Nguyen, H., Hurley, D., Agu, E.: Fact or fiction. Proc. ACM Hum. Comput. Interact. (CSCW) 2, 111 (2018)

27. Lowe, R., Pow, N., Serban, I.V., Pineau, J.: The ubuntu dialogue corpus: a large dataset for research in unstructured multi-turn dialogue systems. In: Proceedings of the SIGDIAL (2015)

28. Munson, S.A., Lee, S.Y., Resnick, P.: Encouraging reading of diverse political viewpoints with a browser widget. In: Proceedings of the ICWSM (2013)

29. Naver News. http://news.naver.com (2019). [Online; Accessed 15 April 2019]

30. Park, S., Kang, S., Chung, S., Song, J.: Newscube: delivering multiple aspects of news to mitigate media bias. In: Proceedings of the CHI, pp. 443–452. ACM (2009)

31. Pascanu, R., Mikolov, T., Bengio, Y.: On the difficulty of training recurrent neural networks. In: Proceedings of the ICML (2013)
32. Reis, J., Benevenuto, F., de Melo, P.V., Prates, R., Kwak, H., An, J.: Breaking the news: first impressions matter on online news. In: Proceedings of the ICWSM (2015)
33. Reuters Institute Digital Report. http://www.digitalnewsreport.org/survey/2016/ (2016). [Online; Accessed 15 April 2019]
34. Riedel, B., Augenstein, I., Spithourakis, G., Riedel, S.: A simple but tough-to-beat baseline for the fake news challenge stance detection task. corr abs/1707.03264 (2017)
35. Rony, M.M.U., Hassan, N., Yousuf, M.: Diving deep into clickbaits: who use them to what extents in which topics with what effects? In: Proceedings of the ASONAM, pp. 232–239. ACM (2017)
36. Samek, W., Wiegand, T., Müller, K.R.: Explainable artificial intelligence: understanding, visualizing and interpreting deep learning models. arXiv preprint arXiv:1708.08296 (2017)
37. Shen, T., Lei, T., Barzilay, R., Jaakkola, T.: Style transfer from non-parallel text by cross-alignment. In: Proceedings of the NIPS, pp. 6830–6841 (2017)
38. Smith, E.J., Fowler, G.L. Jr.: How comprehensible are newspaper headlines? SAGE Publications: Los Angeles (1982)
39. Tai, K.S., Socher, R., Manning, C.D.: Improved semantic representations from tree-structured long short-term memory networks. In: Proceedings of the ACL, vol. 1, pp. 1556–1566 (2015)
40. Talos, C.: Fake news challenge – team SOLAT IN THE SWEN. https://github.com/Cisco-Talos/fnc-1 (2017). [Online; Accessed 15 April 2019]
41. Tannenbaum, P.H.: The effect of headlines on the interpretation of news stories. Journal. Bull. **30**(2), 189–197 (1953)
42. Wang, Z., Hamza, W., Florian, R.: Bilateral multi-perspective matching for natural language sentences. In: Proceedings of the ICJAI, pp. 4144–4150. AAAI Press (2017)
43. Wei, W., Wan, X.: Learning to identify ambiguous and misleading news headlines. In: Proceedings of the IJCAI, pp. 4172–4178. AAAI Press (2017)
44. Yang, Z., Yang, D., Dyer, C., He, X., Smola, A., Hovy, E.: Hierarchical attention networks for document classification. In: Proceedings of the NAACL, pp. 1480–1489 (2016)
45. Yoon, S., Shin, J., Jung, K.: Learning to rank question-answer pairs using hierarchical recurrent encoder with latent topic clustering. In: Proceedings of the NAACL-HLT, vol. 1, pp. 1575–1584 (2018)
46. Zhou, P., Shi, W., Tian, J., Qi, Z., Li, B., Hao, H., Xu, B.: Attention-based bidirectional long short-term memory networks for relation classification. In: Proceedings of the ACL, pp. 207–212 (2016)

An Evolving (Dis)Information Environment – How an Engaging Audience Can Spread Narratives and Shape Perception: A Trident Juncture 2018 Case Study

Katrin Galeano, Rick Galeano, and Nitin Agarwal

Abstract This chapter provides an overview of the evolving YouTube information environment during the NATO Trident Juncture 2018 exercise. The spread of information is no longer solely driven by information actors publishing content on social media, but also by the audience that interacts with it. Engagement features, such as comments and replies, allow an audience to interact with the publisher and other users. This research focuses on the impact that commenters on YouTube have on boosting influence for channels. YouTube channels are able to interact with their audience in the comment section which can be used and abused to spread messages and disinformation. This study focuses specifically on YouTube comments posted around the 2018 Trident Juncture exercise, the largest NATO exercise in recent decades, and identifies how commenters propel video's popularity while potentially shaping human behavior through perception. YouTube is the most popular social media site for video sharing. With that, YouTube channels influence human network behavior by shaping perceptions; simultaneously, commenters on these channels boost search engine results which promulgates higher returns on search engines. Presented is an in-depth analysis of comments and commenters on YouTube channels covering Trident Juncture. Comments by individuals drove both popularity and perception. Additionally, commenters helped in amplifying the messages of the channels. This research reveals effective communication strategies that are often overlooked but highly effective to gain tempo and increase legitimacy in the overall information environment.

Keywords Commenter · Co-commenter · NATO · Information maneuver · Disinformation · Social network analysis · YouTube · Social media engagement · Social hysteria propagation · Information actor analysis · Discourse analysis ·

K. Galeano · R. Galeano · N. Agarwal (✉)
University of Arkansas at Little Rock, Little Rock, AR, USA
e-mail: kkaniagalea@ualr.edu; ragaleano@ualr.edu; nxagarwal@ualr.edu

© Springer Nature Switzerland AG 2020 253
K. Shu et al. (eds.), *Disinformation, Misinformation, and Fake News in Social Media*, Lecture Notes in Social Networks,
https://doi.org/10.1007/978-3-030-42699-6_13

Manipulation and deception analysis · Online deviant behavior · Trident
juncture · Information environment · Algorithmic manipulation

1 Introduction

YouTube is the largest storytelling platform that incorporates videos from across
the globe allowing for freedom of expression, freedom of information, freedom of
opportunity and freedom to belong according to YouTube [1]. The popularity of the
medium led to the creation of programs produced just for this platform. Viewers
have the ability to interact with the content and its publishers by commenting
on videos and replying to comments posted by the channel and other viewers.
Comments may be used for a wide variety of intentions. The commenter may want
to spread a message or take advantage of YouTube's algorithm ranking that will
result in an increase of ranking. We created a workflow that allows us to explore
the larger co-commenter network. A description of this is explained more in the
methodology section.

Specifically, this study focuses on the co-commenter network that was created
when YouTube users commented on channels that published content about a
large NATO exercise called Trident Juncture 2018 [2] (Table 1). The exercise
was conducted in October/November 2018 and remains one of the largest NATO
exercises to integrate land, sea, air, and cyberspace. It simultaneously provided
an overwhelming opportunity for Russian influence to engage in information

Table 1 List of channels used for this research that published Trident Juncture videos between
October 17 and November 20, 2018

YouTube Channels	
NATO	NATO-official
SHAPE NATO	NATO-official
NATO JFC Naples	NATO-official
OTAN	NATO-official
Bundeswehr	NATO-affiliated
alconafter	Anti-NATO
R G D NEWs	Anti-NATO
Youtupe Mania	Anti-NATO
Gung Ho Vids	Anti-NATO
Hoje no Mundo Militar	Anti-NATO
RT	Anti-NATO
RT Deutsch	Anti-NATO
KlagemauerTV	Anti-NATO
Latest News 360	News channel for military enthusiasts
Defense Flash News	News channel for military enthusiasts
Weapons of the World	News channel for military enthusiasts

confrontation. Trident Juncture demonstrated the fusing of twenty nine nations and multiple Partner for Peace nations to operate as a larger coalition alliance, in addition to establishing the foundations for operating in the social media domain through numerous experimentations of new applications as well as analysis of the information environment through internal assessments as well as outreach to academia.

The takeaway from this research is how adversary media campaigns are active in the digital information environment, specifically via YouTube and how disinformation was weaponized via comments on YouTube. We conducted social network analysis to identify the commenter network that propelled video and subsequently channels popularity in the data set. The social network analysis allows for the illumination of hidden networks that are key to disinformation spread and shaping of narrative change. The study shows that despite the misnomer that comments historically appeared irrelevant to nodal influence they actually have a hidden effect of increasing a social media post's influence and thus should be monitored and actioned on just the same as a regular post. Utilizing comments as a means of information maneuver allows for a "back door" that is in plain sight which anyone

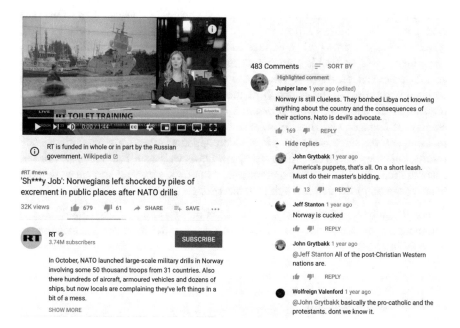

Fig. 1 Screen capture of the RT YouTube channel's video "'Sh***y Job': Norwegians left shocked by piles of excrement in public places after NATO drills" displays comments from viewers and replies

Fig. 2 Commenter and
Co-commenter network with
comments on at least ten
videos

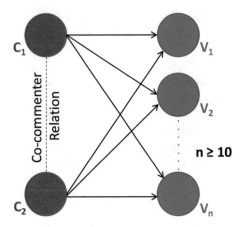

can walk through to influence a network; this coupled with co-commenters amplifies messaging.

Commenter YouTube Network A commenter YouTube network is the elementary network in relation to an overall feed in which comments are made under a post. The example in Fig. 1 shows a post from the Defense Flash News YouTube channel, which at the same time will be defined as the node. The administrator of a channel can either enable or disable comments from viewers in settings; the channel is also able to comment on its own videos and reply to other comments. By receiving comments on videos, the outreach of these videos increases as its being ranked higher [3]. In addition, when a channel replies to comments, it can double the amount of engagement; thus, pushing up the ranking even more. The higher the ranking, the further toward the top a video appears in not only YouTube search results, but Google web and video search results as well as related video sections of specific video pages. Therefore, comments can quickly expand audiences without the need for amplification through paid marketing.

For the purpose of this study, commenters and co-commenters would have to comment on the same videos at least ten times in order to have a connection, as depicted in Fig. 2.

2 Literature Review

YouTube research does not have the same academic depth of published research compared with social media platforms similar in size, functionality, and stature such as Facebook and Twitter. Whereas the aforementioned comparison pertains to the overall global reach and number of users. The derivative of this literature review provides the framework to shape YouTube C_1 and C_2 research for fellow academia using new tools such as YouTubeTracker and others as they are developed. For this research we conducted reviews of both prevailing academic papers as well as proven research that has been validated time and time again, such as literature on social network analysis metrics. Cheng et al. identified "that a power low can fit better than a linear fit" [4]. Therefore, a video can have an increasing growth (if the power is greater than 1), their research went on to identify that only 30% of videos fall into this category which indicates that most videos grow more and more slowly as time passes, also known as the videos active life span. In order to review large amounts of data in networks of this size, research by the RAND Corporation concluded that 'big data tools' were needed to focus on investigation efforts [5]. Hence, the effort partaken for this utilized a number of applications to support investigation of this magnitude with big data.

There is a growing body of researchers that have identified Russian influence as a primary actor in this realm, with recent studies from the RAND Corporation describing how Russian-affiliated accounts add content to user-generated sites, such as YouTube, and "also add fear-mongering commentary to and amplify content produced by others" [6]. While research on comments made on news websites and blogs has been conducted over the years, an in-depth exploration of YouTube commenter and co-commenter networks and YouTube in general, is still fairly contemporary. Studies of social media influence as a whole, provide an overarching acumen such as the 2018 "Russian Social Media Influence: Understanding Russian Propaganda in Eastern Europe" report [6], in which the authors propose a series of options ranging from rule of law, increase in U.S. executive and legislative branches, industry cooperation, and increase funding for research for academia that would allow for development of better tools that identify disinformation on social media. Although all the points are extremely important, the spotlight in reference to academia is paramount due to the overarching speed in which social media is moving. Academia has the digital natives, the tools, and the depth of knowledge to support development of this research and ultimately operationalize it.

With YouTube being a key player in the overall online realm of social communications, it is also the most relevant video sharing platform globally, publishing more than four hundred hours are uploaded every minute [7]. Every day more than one billion users watch over a billion hours of content [8].

What has not been thoroughly researched on YouTube are the comments and co-comments. In his article "Discussions in the comments section: Factors influencing participation and interactivity in online newspapers' reader comments" Dr. Patrick Weber identified that "the bulk of the research on reader comments is conducted in studies on participatory journalism . . . " [9] whereas, social media has had much less research in this field. He went on to demonstrate how comments influence outcomes by readers. This appears to hold true between the media of print and digital and where we have focused on new media (social media) instead of traditional media.

We have defined a comment and a co-comment be connected only if they have both commented on at least ten videos. Sounding somewhat restrictive, it actually generates large networks. Our research identified tens of thousands of nodes and millions of edges thus the creation of a filtered network for this project. As a proof of concept researchers of the project CONTRA: Countering Propaganda by Narration

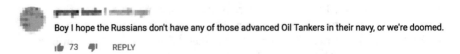

Boy I hope the Russians don't have any of those advanced Oil Tankers in their navy, or we're doomed.

👍 73 👎 !   REPLY

Fig. 3 Screen capture of a YouTube comment about the Norwegian frigate collision

Towards Anti-Radical Awareness [10] have conducted exploratory research on comments of YouTube videos, finding it as a standard feature of web 2.0 features, it allows for general audience participation to the correction of inaccuracies. This is beneficial, if the participation actually provides facts to inaccuracies, but as a matter of conjecture most comments appear to insert hyperbole. Figure 3 below is in reference to a Norwegian military ship that ran aground at the end of exercise Trident Juncture 2018, subsequently it sank; a perfect example of hyperbole about Trident Juncture 2018. Comments of this nature were continuously observed on YouTube channels in order to make a post more influential or to derail a narrative of a message.

3 Methodology

Building on our information environment assessment experience from Trident Juncture 2015, we set out and designed a longitudinal study over a seven-month period. In May 2018, we planned the information environment assessment with NATO subject matter experts and began data collection training. The following 3 months, we conducted two test data collections to prepare for the main data collection during Trident Juncture from October 17 until November 20, 2018.

Data was collected utilizing YouTube API. We also reviewed YouTube content manually on a daily basis during our data collection period which assisted us in the identification of channels to analyze. We chose NATO-owned channels NATO, SHAPE NATO, NATO JFC Naples and OTAN as well as the German delegation's Bundeswehr. In addition, we collected videos published by channels (Latest News 360, YouTupe Mania, alconafter, Hoje no Mundo Militar, Gung Ho Vids, R G D NEWs – Новости, факты, события, RT, RT Deutsch and KlagemauerTV) that countered the NATO narratives as well as military news channels that covered Trident Juncture to appeal to military enthusiasts (Defense Flash News, US Defense News and Weapons of the World). All channels published Trident Juncture 2018-related videos during our data collection time period.

Data was collected utilizing YouTube's API. The following attributes of the videos were obtained: URL of the video, video ID, title of the video, description of the video, number of views, number of likes, number of dislikes, number of comments the video received at the time data was collected, comment text, commenters' unique identifiers, title of the channel that published the video, and the number of subscribers of the channel.

In order to create the commenter network, we multiplied the network by its transpose. These calculations resulted in a square symmetric matrix. This folded network is a commenter-commenter network, where link values are the number of videos the two commenters commented on. For the purpose of this study, co-commenters would have to comment on at least ten of the same videos in order to have a connection.

We then conducted social network analysis of the co-commenter network utilizing the software ORA [11]. Social network analysis allowed us to dive deeper to identify key information actors. Identifying the dark networks that are not normally illuminated upon first review was important in order to identify how these co-commenters collaborated internally within the overall network.

4 Analysis

4.1 Co-commenter Network Analysis

Utilizing this data and analyzing the co-commenter network was not only a process of our overall research, but we found it imperative as we identified that 35,601 users commented on 503 videos that were published by the selected channels during the 34-day timeframe. The comments and co-comments tied together more than 9000 nodes with over 4.4 million edges crawled over this period in 2018 (Table 2).

Folded network data is generally large in size, whereas this date set is extremely large with 35,601 nodes and 47,598,096 edges. The dataset is connected if both C_1 and C_2 commented on at least ten videos or more together. By narrowing it down to at least ten videos, the network was reduced to 583 nodes (which represents the commenters) connected by 5844 edges. We removed 35,018 isolates. The minimum of ten and the maximum of 118 videos that commenters had in common showed a large difference among some commenters. The mean of 14.82 indicated that most commenters commented on 10–14 of the same videos.

We then calculated the density of the network, which is a measure of cohesion and equals the ratio of the number of actual ties to possible ties [12]. More connected networks have higher density scores and are more cohesive, whereas lesser connected networks have lower scores and are less cohesive [13]. The total density of the network is 0.0172.

To gain an insight of the key information actors in the network, we performed centrality calculations. Three commenters stood out repeatedly when measures of centrality were calculated (Table 3). These actors consistently ranked in the same

Table 2 Co-commenter network data

Nodes:	583	Minimum links:	10
Links:	5844	Maximum links:	118
Total density:	0.0172	Mean:	14.82

Table 3 Centrality measurement results identifying the top three commenters within the co-commenter network

Ranking	Total Degree	Betweenness	Closeness	Eigenvector
1	0.148	0.605	0.603	0.587
2	0.071	0.117	0.445	0.416
3	0.071	0.1	0.444	0.394

Fig. 4 Screen capture of YouTube commenter Juniper Lane profile located

order for total-degree centrality, betweenness centrality, closeness centrality, and Eigenvector centrality.

Betweenness centrality score for actor 1 revealed a score of .605 which indicates that in certain respects this actor is very centralized. In this case, the network revolves around a broker in a position to control the flow of information through the network. This actor also displays very inorganic behavior, indicative of a bot. For example, this account was created only weeks before the start of the exercise (September 13, 2018). The use of a cat image for a profile and the name of Juniper Lane does not reveal any personally identifiable information. The account has not published any videos and just as recently as June 23, 2019, gained all of its eight subscribers (Fig. 4).

A robotic actor (a botnet) would be able to process and disseminate the information much more rapidly than a human. Whereas, actor 2 and 3 have betweenness scores of .117 and .100, respectively. Meaning that these agents are much more decentralized; this was our first indicator that they are real people. They appeared to have active accounts, that spanned several years, in addition the other two agents could be seen with historical comments on other YouTube channels that made sense.

The co-commenter network created for this study displayed in Fig. 5 visualizes the network's most central actors. The nodes are colored based on their closeness centrality on a hue color scale with red being the most central and blue being the least central. The node size was also adjusted based on centrality. The larger the node, the higher its centrality. The links are colored by value. The greater the number of shared videos two commenters commented on, the darker the link. In addition, we also increased the width of the links based on value. These results echoed the aforementioned betweenness centrality in which the agents remained in the same ranking order (agent 1–3). This further corroborated that of our previous evidence that actor 1 is inorganic and is displayed in this sociogram.

As seen in Table 2, actor 1 ranked as the top node for all centrality measures and is the most connected node within the network. The network graph above highlights this node with its 564 edges. We further analyzed this node and the other two nodes

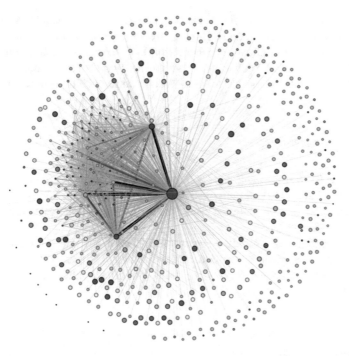

Fig. 5 Co-Commenter Network prominently displaying the node with the most links (564 edges). The nodes are colored based on their closeness centrality on a hue color scale with red being the most central and blue being the least central. The node size was also adjusted based on centrality. The larger the node, the higher its centrality. The links are colored by value. The greater the number of shared videos two commenters commented on, the darker the link. In addition, we also increased the width of the links based on value

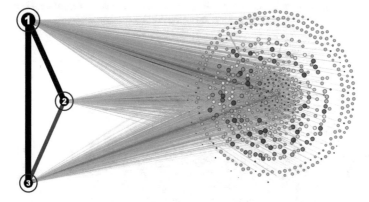

Fig. 6 Ego network of the top three commenters. The nodes are colored based on their closeness centrality on a hue color scale with red being the most central and blue being the least central. The node size was also adjusted based on centrality. The larger the node, the higher its centrality. We have numbered the top three central nodes. The links are colored by value. The greater the number of shared videos two commenters commented on, the darker the link. In addition, we also increased the width of the links based on value. All three of these accounts are both anti-NATO and anti-US

that were most centralized which stood out and compared to the rest. As referenced in the methodology about ego networks, we sought out to explore this. When we created a Sphere of Influence (ego network) of these three nodes (Fig. 6), they directly connected to 575 other nodes with a total of 5834 links. Additionally, these nodes were directly connected to 99.14% of all nodes within the network.

4.2 Topic Analysis

All of the three most central co-commenters commented on channels with anti-Western and anti-NATO narratives. These actors added to the information environment and were aligned with the messages spread by the channels. Table 4 shows sample comments.

These comments provide a small example of the many comments these users posted. In order to explore these comments further, we created a word cloud for all comments these three users published within the co-commenter network.

Table 4 Sample comments from the most central three actors within the network

Commenter	Comment
Actor 1	Norway is still clueless. They bombed Libya not knowing anything about the country and the consequences of their actions. Nato is devil's advocate.
Actor 2	Such hypocrites. ....as if the USA are any better than Russia. Disgusting warmongers. They are endangering us all with their aggressive warmongering ways.
Actor 3	The Norwegian beer was just too strong. Norway better align with Russia. More spiritually alike and can handle alcohol. And Russian soldiers are polite and good guests. This is all an expression of the driving root mindset.

Table 5 Word cloud for top ranking co-commenter

Actor 1

(continued)

Table 5 (continued)

Actor 2

Actor 3

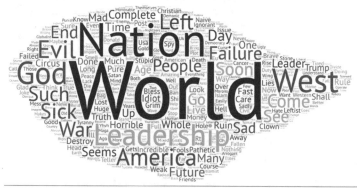

The word cloud shows the messaging that was highlighting topics on YouTube. The relevancy of these particular words is relative to taking away the focus of Russia and refocusing on trending news (at the time). These word clouds are rank ordered based upon the top three influential nodes

The relative size of the words in the word clouds displayed in Table 5 represents the frequency of the words used in the comments. The most central commenter frequently wrote about war, crisis, meddling, and migration.

5   Conclusions and Future Work

A commenter YouTube network is the elementary network in relation to an overall feed in which comments are made under a post. By allowing for comments on posts the outreach increases and audiences can rapidly expand without the need for amplification through paid marketing. But, combining this tactic of commenting, co-commenting, and paid marketing drives channels through the roof with influence.

Our data has set the foundation for how we will view these entities and their relationships in the future. We visualized how YouTube networks between C1 and C2 are connected and the studies/theories that demonstrate the accepted and proven practice of filtering and exploring large data sets. Future studies will include a detailed cyber forensics analysis of the most central nodes within the co-commenter network. We will also further study user behavior. Actor 1, for instance, only commented on each video once whereas the other two actors commented on some videos multiple times. Consistent misspellings, the use of British English and the lack of interaction with other commenters who replied or liked the actor's comments also leads us to believe that this actor is not human.

Acknowledgment This research is funded in part by the U.S. National Science Foundation (OIA-1920920, IIS-1636933, ACI-1429160, and IIS-1110868), U.S. Office of Naval Research (N00014-10-1-0091, N00014-14-1-0489, N00014-15-P-1187, N00014-16-1-2016, N00014-16-1-2412, N00014-17-1-2605, N00014-17-1-2675, N00014-19-1-2336), U.S. Air Force Research Lab, U.S. Army Research Office (W911NF-16-1-0189), U.S. Defense Advanced Research Projects Agency (W31P4Q-17-C-0059), Arkansas Research Alliance, and the Jerry L. Maulden/Entergy Endowment at the University of Arkansas at Little Rock. Any opinions, findings, and conclusions or recommendations expressed in this material are those of the authors and do not necessarily reflect the views of the funding organizations. The researchers gratefully acknowledge the support.

References

1. YouTube. About YouTube – YouTube. https://www.youtube.com/yt/about/
2. NATO – Exercise Trident Juncture 2018. In: NATO. https://www.nato.int/cps/en/natohq/157833.htm (2018)
3. Stelzner, M.: YouTube ranking: how to get more views on YouTube (2019). https://www.socialmediaexaminer.com/youtube-ranking-how-to-get-more-views-on-youtube-sean-cannell/
4. Cheng, X., Dale, C., Liu, J.: Statistics and social network of YouTube videos. In: 2008 16th International Workshop on Quality of Service, pp. 229–238 (2008)
5. Bodine-Baron, E., Helmus, T., Radin, A., Treyger, E.: Countering Russian social media influence. https://doi.org/10.7249/RR2740 (2018)
6. Helmus, T.C., Baron, E.B., Radin, A.A., Magnuson, M., Mendelsohn, J.: Russian Social Media Influence: Understanding Russian Propaganda in Eastern Europe. Santa Monica, CA: RAND Corporation. https://doi.org/10.7249/RR2237 (2018)
7. Tran, K.: Viewers find objectionable content on YouTube kids. https://www.businessinsider.com/viewers-find-objectionable-content-on-youtube-kids-2017-11 (2017)
8. YouTube. YouTube for press. https://www.youtube.com/yt/about/press/
9. Weber, P.: Discussions in the comments section: factors influencing participation and interactivity in online newspapers' reader comments. New Media Soc. **16**, 941–957 (2014)
10. Ernst, J., Schmitt, J.B., Rieger, D., Beier, A.K., Vorderer, P., Bente, G., Roth, H.: Hate beneath the counter speech? A qualitative content analysis of user comments on YouTube related to counter speech videos. J. Deradicalization. **10**, 1–49 (2017)
11. ORA-LITE: Software | CASOS. http://www.casos.cs.cmu.edu/projects/ora/software.php
12. Everton, S.: Disrupting Dark Networks. Cambridge University Press, New York (2012)

13. Although the density metric is useful, it has its limitations since this density score is sensitive to the size of the network. Larger networks have lower density scores than smaller networks; this metric should only be used when comparing networks of similar size

Blockchain Technology-Based Solutions to Fight Misinformation: A Survey

Karen Watts DiCicco and Nitin Agarwal

Abstract Blockchain has been around since 2009, but it isn't till the last few years that organizations have been looking into using blockchain for other applications than cryptocurrency. One of these areas is using blockchain for addressing the problem of misinformation. These emerging solutions that are being used to fight and prevent misinformation range from validating news articles, images, videos, and even entire social media platforms. Each blockchain technology-based solution has pros and cons and adopt different approaches on how they aim to prevent and fight misinformation in social media.

Keywords Blockchain · Misinformation · Fake news · Social media · Survey

1 Introduction

Mostly when one thinks of Blockchain, the first thing that may come to their mind is cryptocurrency. Today blockchain is used for more than cryptocurrency. Organizations are learning and researching about how they can use this emerging technology to solve their use cases and for building their platforms on. Blockchain is being used to solve food traceability, voting, verifying resumes, supply chains, and many other areas. One area that is emerging is using blockchain to fight and prevent misinformation. Misinformation is a growing problem on social media. Misinformation can be in the form of text, news articles, images and videos. This is a big problem when you think that factor in that about two-thirds of American adults (68%) say they at least occasionally get news on social media. About four-in-ten Americans (43%) get news on Facebook. The next most commonly used site for news is YouTube, with 21% getting news there, followed by Twitter at 12%. Smaller portions of Americans (8% or fewer) get news from other social networks

K. W. DiCicco · N. Agarwal (✉)
Department of Information Science, University of Arkansas – Little Rock, Little Rock, AR, USA
e-mail: klcottrill@ualr.edu; nxagarwal@ualr.edu

© Springer Nature Switzerland AG 2020
K. Shu et al. (eds.), *Disinformation, Misinformation, and Fake News in Social Media*, Lecture Notes in Social Networks,
https://doi.org/10.1007/978-3-030-42699-6_14

like Instagram, LinkedIn or Snapchat [1]. To be able to solve these problems, many organizations are researching and developing blockchain solutions to help fight and prevent misinformation, and this number increases daily as new platforms are being developed. These platforms range from verifying news articles, social media content, images, and videos. Using blockchain technology to fight misinformation is emerging and the solutions that are currently available are still in research, prototype, and beta testing stages.

In this paper, we look into blockchain technology Sect. 2, to get a brief overview of what is blockchain Sect. 2.1, how it works and how organizations are using this technology for other applications besides cryptocurrency. Section 2.2 explains why use it for misinformation to verify content, images, videos, and social media. Section 3 discusses what platforms are out there and the pros and cons of each that mention they are using blockchain to fight and prevent misinformation. Section 4 discusses other blockchain platforms that are on blockchain but are not stating they are using blockchain to fight and prevent misinformation. Section 5 explains opportunities that are available for developers to be involved. Section 6 discusses ethical and policy concerns of blockchain and laws and regulations. Section 7 provides challenges of using blockchain for misinformation. Last, Sect. 8 provides suggestions that should be used to make blockchain a successful solution to fight and prevent misinformation.

2 Blockchain

2.1 What Is Blockchain

Blockchain is a collection of blocks of data. These blocks of data contain the information that a user wants to be stored on the blockchain and the hash of the previous block (Fig. 1). These blocks are spread across different computers that are linked in a peer-to-peer network. With the data being spread across nodes in the peer-to-peer network, this makes it hard to be able to hack. The data that is on the chain is verified by checking all the transactions in the chain. What was originally created for Bitcoin is now used for many other use cases, including misinformation. Advantages that blockchain has over a database is that it provides a proof of ownership and data can't be deleted from the chain, creating a transparent, traceability, and immutable record.

For the organizations that are researching and creating solutions, blockchain is a way for them to make the content transparent, traceable, and immutable. For example, if we want to verify an article, there are two ways this can be added to the blockchain. One is that all the content is uploaded to the chain or you can upload the file information to the chain. Depending on the use case, will depend on which option would be used. If a user decides to upload all of the content to be stored on the chain, this option would take up more storage, where the second option is just

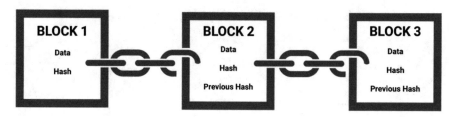

Fig. 1 Concept of Blockchain

the file information, this option would take up less storage. If a user shares the file information with another user, they would be able to verify that file is the original file on the blockchain. Using either one of these options would be able to provide proof that the file is the original and the author is the original author. Where how it is now, authors can't verify their content is there original work.

Blockchain is being used in different industries from agriculture, supply chains, and absentee voting. In agriculture Wal-Mart has partnered with IBM Food Trust for food traceability. The first produce that Wal-Mart has mandated is leafy greens. All leafy green suppliers had until September 2019 to be on Wal-Marts blockchain. Where it once took a little over 6 days to trace a mango, they are able now to trace where leafy greens came from in less than 3 s. J.B. Hunt, Tyson, and Wal-Mart are looking to use blockchain for their supply chains. Currently, when shipping from the supplier to the customer, the bill of lading is a paper form. By using blockchain, they would be able to make everything digital and be able to verify shipments. If a shipment is a refrigerated shipment, they can track this with sensors and if there is an issue it would terminate the smart contract and store this information on the blockchain along with the data from the shipping forms. This could prevent disputes that the industry currently has on shipments. In the last year, West Virginia and Utah tested a small pilot with the company Voatz (https://voatz.com) for absentee ballots. This is making it possible for voters to vote from a mobile device and make it immutable using blockchain.

2.2 Why Use It for Misinformation

Misinformation is a growing problem in news and social media outlets and is also easily spread by sharing altered content, photos, and videos. Misinformation has affected society from political campaigns, vaccinations, pesticides and GMOs in Agriculture, and many other areas. Users can set-up fake accounts and spread fake news for their agenda. One example that had everyone's attention about fake news was the presidential election in 2016. According to Truepic [2] photos and videos are acceptable to being editing to add, remove, or modify objects, changing the location information embedded in the photo or video, to make it appear as if were captured somewhere else, metadata manipulation, and using artificial intelligence to

Fig. 2 Blockchain Solutions that are being developed to help fight fake news

fabricate, including deepfakes. Platforms such as Facebook, Twitter, and YouTube are using algorithms, machine learning, and artificial intelligence, to find fake news and take it down from their platforms. Unfortunately, this is not enough, these solutions are not detecting all the fake news that is circulating, and content that has already done damage before it is detected. Currently, there is not a way to verify content by users, images, and videos. A user can go on the internet read an article that may not be the original article and have no idea that what they are reading isn't the truth. Images and videos are also being altered and users are swayed by what they see. By developing a blockchain solution (distribute ledger technology), companies are trying to prevent fake news by verifying content and images as they are created. This would allow users to know who the original author is and to see if what they are viewing has been altered. Figure 2 provides an illustration of the evolution of blockchain based solutions in this space. Next section provides details about each of these technologies and a comparative analysis.

3 What Platforms Are Out There and the Pros and Cons of Each

In the last 4 years platforms have been emerging to use blockchain to fight fake news. Each platform is modeled different on what they offer (Fig. 3). Out of all the platforms researched, currently only one verifies users, two offer plug-ins, and one offers an API to use for developers building applications.

Blockchain Platforms

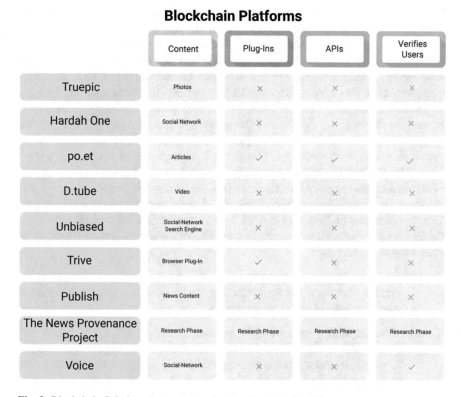

	Content	Plug-Ins	APIs	Verifies Users
Truepic	Photos	×	×	×
Hardah One	Social Network	×	×	×
po.et	Articles	✓	✓	✓
D.tube	Video	×	×	×
Unbiased	Social-Network Search Engine	×	×	×
Trive	Browser Plug-In	✓	×	×
Publish	News Content	×	×	×
The News Provenance Project	Research Phase	Research Phase	Research Phase	Research Phase
Voice	Social-Network	×	×	✓

Fig. 3 Blockchain Solutions that are being developed to help fight fake news

3.1 Truepic

Truepic [2] is a photo and video verification platform that has sponsored with Jewelers Mutual Insurance Company, Qualcomm, Ask Me Anything, US Department of State, UNCDF, and Credibly. This platform was founded in 2015 and is growing platform with three different products and has an iOS and Android App. Their platform is based on the mission that they want to fight against disinformation by restoring trust in visual media, they claim to do their part in defending democracy again manipulation through visual deception. Besides blockchain, their platform uses computer vision, and AI for photo verification. Truepic uses a foundational technology called Controlled Capture. This technology verifies the origin, pixel contents, and metadata, as soon as the user hits the capture button. Every photo that is taken with this technology is passed through an unbroken chain of custody; from the time the capture button is pressed to the time the image is shared with the recipient. Every link in the chain of custody is logged and can be verified at any point later. Each photo and video has an unique cryptographic signature and is written to the blockchain, creating an immutable record in the distributed public ledger [2]. Truepic App is available for iOS and Android devices. This application allows anyone user to capture and share photos and videos that have verifiable origin, pixel contents, and metadata. From the moment a user presses the capture button, Truepic's Controlled Capture pipeline receives the data directly from the image sensor, along with data from other onboard sensors such as GPS, It computes a cryptographic signature – like a digital fingerprint – for the image, and encrypts all this information for transmission to the Truepic's servers. A unique identifier is created for the image. This initiates the image's "chain of custody". The original image, along with its metadata, cryptographic signature, and the results of the verification tests are stored in the Truepic vault in perpetuity. A verification page is created for the image, allowing recipients to view the original image and the results of the verification tests. An immutable record, containing the image's cryptographic signature, is written to a public distributed ledger that is neither controlled by Truepic nor hosted on Truepic's servers. This closes the chain of custody [2]. Out of the all the platforms that are currently available, this platform is the furthest along in development and usage.

Pros of Truepic: When a user captures a photo, the metadata and GPS is stored with the photo to allow for verification and traceability. The application has been very helpful for insurance claims.

Cons of Truepic: Only works with iOS and Android devices. This will help verify images taken with these devices but not with cameras and video cameras.
To use their application, a user has to have internet connection. This can be a huge problem for users that live in Rural areas that have little to no connection and also for users that may be in a building that has poor connection.
You can still stage a photo or take a photo of an existing photo and the application will verify the photo.
There is no verification process that the user that signs up to the platform is a real person.

3.2 Po.et

The po.et [3] foundation was founded in 2017 and is currently in phase 1 of development. Po.et is building an open, universal ledger that records immutable and timestamped information about your creative content that uses their open protocols designed for interoperability with current industry standards in media and publishing. The platform exists to help content creators establish immutable, provable layers of value to help create a better web. The three problems po.et is addressing is ownership, utilization, and history [3]. This platform is great for anyone that is wanting to use their service to validate they are the owner or authors. Po.et allows content creators to integrate with plug-ins from content management systems. The first plugin was for WordPress, and has now expanded to Drupal and Joomla!

The po.et network focuses on the following:

Attribution
The foundation of Po.et is attribution. We enable multiple new ways to reference that content through the entire network to help establish validated claims such as ownership or authorship.

Discovery
The Po.et Network builds a set of ubiquitous information based on standard protocols to lower the friction in communicating the value of our content and how it can be unlocked.

Monetization
With the combination of verifiable reputation and on-chain discovery, the Po.et Network expands the options for monetizing content in a safe, controlled way. Both public and private marketplaces can be enabled.

Reputation
Everyone can see the actions taken on the Po.et Network and establish a history of certain behaviors by entities. Each piece of new information helps us be sure of who is safe with which to engage [3].

Pros of po.et: For content creators this is a great way to validate ownership of their work.
It's easy to connect to the system using one of the content management solutions plugins.

Cons of po.et: Currently the platform only supports text, but is working on other supported file types (photos, videos, PDFs)
It can be hard to integrate with the platform if you not using one of the available plugin platforms, and you're not a developer or have someone that can integrate with their platform through the API.
There is no verification process that the user that signs up to the platform is a real person.

3.3 New York Times News Provenance Project

The News Provenance Project [4] is exploring new ways for publishers to help fight misinformation. The first phase of this project is to do technical and user research and turn this research into a proof of concept, focused on photojournalism. The first phase of this project will end late 2019. The News Provenance Project will be using IBM Hyperledger for their proof of concept and will be collaboration with IBM Garage. The project is being spearheaded by The New York Times's Research and Development team, which is made up of technologists and journalists who explore the potential of emerging technologies for journalism [4].

In developing potential solutions, The News Provenance Project has several components:

- Conducting user experience research to understand what kinds of signals and indicators are intuitive, useful and relevant to people in the context of their daily routines around news.
- Implementing a technical proof of concept exploring the potential of blockchain, in order to understand how some of its attributes— immutability and decentralization in particular— might be used to guard against manipulation and enhance users' trust in the news material they come across.
- Building a working group to collaborate on future experimentation.
- Identifying a sustainable model for adoption and participation among publishers and platforms of any proven outcomes [4].

What they hope to learn:

- Could information about a photo's digital history help people better understand the way it is produced and published?
- How much information might be helpful or necessary in sourcing a photo shared outside of its published context?
- What kinds of metadata — for example, the time and place the photo was captured, the original publisher and caption, the photo's revision history— might be important to include or prioritize?
- How helpful might a symbol or watermark be in establishing credibility?
- How might access to photo metadata change how audiences perceive photos that don't have metadata? [4]

Pros of The News Provenance Project: No pros at this time, this platform is in research phase.

Cons of The News Provenance Project: No cons at this time, this platform is in research phase.

3.4 Voice

Voice [5] is a brand-new social network that launched this year that is being developed by Block.one on top of the EOS blockchain. Currently the application is available through beta access. The Voice platform uses a special authentication system to make sure very user on Voice is an actual person. A user can create content and make their post go live and earn Voice tokens. These tokens can be used to put your content on top of other content for users to see. The user is in theory paying for their content to be placed on top like an advertising model. Where it gets interesting is that if a user raises their Voice content above your content, the user gets some of its tokens back plus some extra voice [5].

Pros of Voice: Verifies the user's identity on sign-up, to help prevent fake accounts.

Cons of Voice: All content will not be shown equally, the users will be using tokens to get their content seen above others.

3.5 D.Tube

D.Tube [6] was founded in 2017 and is a decentralized version of YouTube that enables video content producers to get paid in cryptocurrency. The D.Tube platform was built on top of the Steem Blockchain and the IPFS peer-to-peer network. This platform is modeled after YouTube, but unlike YouTube, when you share or comment on a video a user can earn cryptocurrency, the platform cannot sensor videos, this is left up to the users by voting on the videos, there is no algorithms that control the visibility of videos, and there are no ads on the platform. Users earn tokens by posting content, sharing links, and voting on the content. D.Tube was created as a new type of video sharing platform designed to offer a solution to these issues: Re-create trust with a censorship-free decentralized hosting infrastructure, fully transparent and open source code and no collection of personal data. Community-powered moderation: Content's popularity, moderation and classification is determined by user's upvotes, downvotes and tags without algorithmic alteration. A token model to reward all users: A "social blockchain" mechanism distributes cryptocurrency token rewards to all users (creators, influencers, viewers) for their social contributions (post, vote, tag) [7]. With D.Tube being a self-governed platform, the platform has leaders in place. These leaders are voted by the users. Each user is allowed to vote for 5 leaders. The top 10 leaders on the leaderboard are in charge of producing new blocks and securing the infrastructure.

Pros of D.Tube: No algorithms so all content can be shown equally.

Cons of D.Tube: The platform cannot sensor videos, only users can, if the leaders were on the same propaganda agenda.
There is no verification process that the user that signs up to the platform is a real person.

3.6 Unbiased

Unbiased [8] was founded in 2017 and is currently in beta testing. This platform is more than a social platform, they also have a search engine product. Both the social platform and search engine were developed to fight fake news by using crowdsourcing with blockchain, machine learning, and artificial intelligence. Unbiased collects the data from the search engine by using both APIs and web scraping. Unbiased platforms goal is to address human bias and quality issues, fake news, and data integrity. Unbiased is determined to present facts by bringing everyone's opinion to one platform and present the aggregated opinion in an interactive way. They will categorize information based on source and also an advanced AI algorithm to validate the information. They will also use digital token incentives to generate ore quality, trusted data and also to crowd-source growth". The social platform is built for a user to create and share content while connecting with friends and followers, while earing cryptocurrency. Where with the search engine a user can search a topic and get insights on the topic they searched, while earning cryptocurrency for sharing the experience and helping the community [8].

Pros of Unbiased: The platform will have a search engine to search other data that was not created on their platform.

Cons of Unbiased: There is no verification process that the user that signs up to the platform is a real person.

3.7 Publish Protocol

Publish Protocol [9] conceptualized in 2018, that uses the Ethereum blockchain platform. Publish Protocol aims to restore readership and secure financial sustainability for publishers [9]. There are three different roles for Publish, these are consumers, producers, and community editors. Consumers can earn tokens by visiting a site, reading articles, leaving comments, upvoting and downvoting content, sharing content, and staking tokens. They can also earn tokens by pointing out errors in published articles. Producers are users that produce the news content, but to submit a news item, producers must stake a certain number of tokens. The two ways they can receive news tokens if an article is published and is upvoted by consumers. Community editors earn tokens by editing the content. You have to apply to become a community editor the person has to meet a minimum set of criteria [10].

Pros of Publish Protocol: By using the platform, you are verifying the news content.
Users can earn tokens, by contributing to sharing, upvoting, and downvoting.

Cons of Publish Protocol: You have to have tokens to be able to publish your content.
There is no verification process that the user that signs up to the platform is a real person.

3.8 Trive

Trive [11] was founded in 2017, and instead of being a platform, this solution is a browser extension plugin, built on Ethereum. The Trive platform is based on spending time and resources appraising news and information for truth value. Users can use the service for $1 a month. Thrive works as a scoring platform that rates each site and when a user uses the browser plugin, a site with a low score will alert them if the site they are on is an unreliable news source [11].

Trive has 5 types of "players":

- Consumers who consume the news and send stories of interest to the marketplace for Curators to find and research;
- Curators who bid on stories and publish lists of article claims with an incentive to maximize profit, reduce research costs and deliver quality;
- Researches who are incentivized to find and document convincing true data quickly and efficiently;
- Verifiers who verify the supporting evidence collected by the Researchers above, and are awarded if/when the Researcher's err;
- Witnesses who verify research and participate in the truth scoring process, earning a small fee and enjoying the truth discovery process [12].

Each of these roles have a set of incentives that maximizes the search for the truth and minimizes gaming [12].

Pros of Trive: The way the platform verifies content through the players, has several people verifying the content.
The browser plugin can hide stories that are below a certain score if the site has been verified.

Cons of Trive: Works only on sites that have been rated.
Doesn't work on social media posts.

3.9   Hardah One

Hardah One [13], a social browser platform, development began in 2016 and is still in development. What makes their solution unique, is that their app will create a bridge between social networks: Facebook, Instagram, Google, Twitter, LinkedIn, YouTube, Netflix, ... Like other social solutions, their app will not be using filter bubbles and they will be working closely with journalists. Hardah One is also anti fake-news by using blockchain and deep learning to trace flagged fake news and reduce the spread of fake news [13].

The challenges they are solving are:

- Lack of Interoperability
- Filter Bubbles + Fake News
- Social Media Business Models

Pros of Hardah One: The app is a multimedia platform in the form of circles, allowing users to create, share [13].

Cons of Hardah One: At this time, there is not a lot of information on the development of this project to currently see what the cons are of the platform.

4 Other Blockchain Platforms

There are many more emerging social media and news platforms that are built on blockchain. These platforms run like regular social media platforms but are designed were users can earn cryptocurrency for their content. Users also use these platforms to avoid having their content censored and because there are no algorithms used on their content. These platforms however do not talk about preventing or fighting fake news.

Other Blockchain Platforms:

SocialX [14] is a social media platform that is similar to Facebook and Instagram where users can share photo and video content. This platform allows user to earn cryptocurrency for publishing, sharing, and liking content.

Steemit [15] is a website that is powered from Steem blockchain and Steem cryptocurrency. Their platform is based on that users of the platform should receive benefits and rewards for the contribution to the platform.

Appics [16] is a reward-based social media application that has content in varying categories. Users are rewarded with cryptocurrency for their contribution.

PiePie [17] is an iOS and Android application that allows users to share videos and automatically earn cryptocurrency.

Minds [18] is an open source platform that allows for Internet freedom. Users are paid in crypto for their contributions to the community.

Yours [19] is a platform where users create content with a free preview and then they put the rest of the content behind a pay wall. The users are able to set the price they want for other users to view their content.

With any new emerging technology, you have technology that succeeds and technology that fails. Three platforms that were designed to fight misinformation on blockchain and no longer around are Truthem, 4Facts, and Userfeeds.

5 Opportunities

Creating and having a solution that will work to fight and prevent misinformation will not be successful from just one person. To be able to solve the problem everyone will have to collaborate. One platform that offers an API to be able to integrate with their platforms is po.et. Developers can integrate their projects through their API. Integration content is available on their GitHub for elixir, frost JavaScript, C#, Frost-Ruby, and Frost-PHP. This platform allows for collaboration and integration with their platform. The more users that connect and use the platform the more successful it can become to helping fight and prevent misinformation.

Steem allows developers to develop on their blockchain platform for free. Steem has near-instant fee-less transactions and its built-in content specific primitives make building an engaging and functional blockchain-powered application easier than ever [20]. D.Tube is one of many applications that is built on the Steem blockchain.

Using blockchain to verify content can help prevent misinformation. Content can be changed and by using blockchain it prevents the original content from being changed and users can verify the origin on the article, image, or videos. To verity authors on blockchain, platforms can use this technology and store certifications and credentials of authors to verify they are trusted authors and publishers can check this before they decide to publish an article. Depending on how a platform is set-up you can't prevent someone from putting fake news on blockchain, this can still happen. On some platforms if a user generates fake news, users vote on the content to give it a ranking. This would push unreliable news to the bottom of a newsfeed but it's still there since you can't remove content. Clearly there are limitations to existing approaches, which present opportunities for researchers like us and others to improve existing solutions or develop new ones and open doors for innovation for researchers and practitioners alike.

6   Ethical and Policy Concerns/Issues

With any new emerging technology, laws and regulations have to catch up to the technology. Many states had bills in legislation in 2018 for blockchain. However, these bills were for defining nodes and running nodes, appointing blockchain working groups, blockchain task forces, concerns for using blockchain for state records, initiative to implement policies, and authorizing smart contracts. With this just being the beginning of regulations, there are no regulations for using blockchain to fight and prevent misinformation and if it will be valid for using in legal issues. Another issue is that some of the platforms mentioned stated that they don't censor data and have no algorithms. However, some of these platforms are using users to vote and score the content published, which is censoring data, which contradicts the statement that they are not censoring data.

7   Challenges

Using blockchain to help prevent and fight misinformation is still emerging. With all emerging technologies there are going to be challenges to overcome. The cost to develop these platforms is expensive and many of them have raised the start-up capital, but how does the platform continue to support itself after the capital runs out. Data also can't be deleted from the blockchain and storing the data will become more expensive as the data collects on the chain. Each platform has different goals and different use cases, this makes the user have to use different platforms to verify anything from photos, videos, to news articles. For users that are not wanting their content censored, some of the platforms, for example D.Tube, says it does not censor the content. However, this may actually not be the case. It may not be the platform censoring the content, but the users that will censor the content by ranking content. Blockchain alone won't completely prevent and fight fake news. This technology will need to be paired with machine learning and artificial intelligence to make the most out of the technology. All of the platform mentioned are still in the beginning stages and everyone is wondering if blockchain will be the solution or is it just hype to fight and prevent misinformation.

8 Looking Ahead

With blockchain technology emerging, the sky is the limit for organizations to use this technology to help fight and prevent misinformation. Out of all the companies one that has come the furthest in development is Truepic. With each company focusing on different aspects of misinformation there is still room for growth. To make any solution successful there has to be collaboration. Solutions also need to be able to be used from different applications via plug-ins and APIs. The more users that can use a solution across applications the more content, images, and videos can be verified if they are authentic or not. However, if a platform doesn't verify users, how will this prevent fake users to start trying to create fake content. Any solution will need to be able to verify a user is who they say they are. Will blockchain be the solution to fight and prevent misinformation, it is still too early for an answer, but the outlook is looking positive.

Acknowledgments This research is funded in part by the U.S. National Science Foundation (OIA-1920920, IIS-1636933, ACI-1429160, and IIS-1110868), U.S. Office of Naval Research (N00014-10-1-0091, N00014-14-1-0489, N00014-15-P-1187, N00014-16-1-2016, N00014-16-1-2412, N00014-17-1-2605, N00014-17-1-2675, N00014-19-1-2336), U.S. Air Force Research Lab, U.S. Army Research Office (W911NF-16-1-0189), U.S. Defense Advanced Research Projects Agency (W31P4Q-17-C-0059), Arkansas Research Alliance, and the Jerry L. Maulden/Entergy Endowment at the University of Arkansas at Little Rock. Any opinions, findings, and conclusions or recommendations expressed in this material are those of the authors and do not necessarily reflect the views of the funding organizations. The researchers gratefully acknowledge the support.

References

1. Shearer, E., Matsa, K.E.: News use across social media platforms 2018. Pew Research Center's Journalism Project. https://www.journalism.org/2018/09/10/news-use-across-social-media-platforms-2018/ (2019)
2. Truepic. https://truepic.com
3. Po.et – The decentralized protocol for content ownership, discovery and monetization of media. https://www.po.et/

4. The News Provenance Project. The News Provenance Project. https://www.newsprovenanceproject.com/
5. Social as it should be. Voice. https://voice.com/
6. D.Tube. https://d.tube/
7. Turning the tables in the social media industry a new model where users vote on videos to reward all contributors: creators|curators|influencers|viewers. D.Tube, 2019. https://token.d.tube/whitepaper.pdf
8. An incentivized social platform built using blockchain and a search engine presenting the real opinion. Fighting fake news and misinformation on the internet using technology|blockchain based circular economy & rewards for content creators. https://unbiased.cc/
9. Editor, PUBLISH. How the PUBLISHprotocol reward system incentivizes community participants. Medium, PUBLISH Newsroom. https://medium.com/publishprotocol/news-token-reward-system-how-does-it-work-7c9313176c17 (2019)
10. Publish. PUBLISHprotocol. PUBLISHprotocol. https://www.publishprotocol.io/#m1
11. Reale, P.: How it works – trive – stop fake news. Trive. https://trive.news/how-trive-works/
12. Mondrus, D., et al.: Trive whitepaper. Trive. https://trive.news/wp-content/uploads/2018/02/Whitepaper.0.2.6x.pdf
13. Hardah. Hardah one project: the blockchain social browser. Hardah one project|the blockchain social browser. https://hardah-one.com/
14. SocialX. https://socialx.network
15. Stemmit. https://steemit.com/faq.html#What_is_Steemit_com
16. Appics. https://appics.com/#
17. PiePie. https://piepieapp.com/
18. Minds. https://www.minds.com/
19. Yours. https://www.yours.org/
20. More transactions than bitcoin and ethereum combined, which means it can easily handle every transaction your app generates. Steem. https://steem.com/developers/

Index

Printed in the United States
by Baker & Taylor Publisher Services